Intention, Common Ground and the Egocentric Speaker-Hearer

Mouton Series in Pragmatics   4

Editor
Istvan Kecskes

Mouton de Gruyter
Berlin   ·   New York

Intention, Common Ground and the Egocentric Speaker-Hearer

edited by

Istvan Kecskes
Jacob Mey

Mouton de Gruyter
Berlin · New York

Mouton de Gruyter (formerly Mouton, The Hague)
is a Division of Walter de Gruyter GmbH & Co. KG, Berlin.

♾ Printed on acid-free paper which falls within the guidelines
of the ANSI to ensure permanence and durability.

Library of Congress Cataloging-in-Publication Data

Intention, common ground and the egocentric speaker-hearer / edited
by Istvan Kecskes and Jacob Mey.
     p. cm. − (Mouton series in pragmatics ; 4)
  Includes bibliographical references and index.
  ISBN-13: 978-3-11-020606-7 (cloth : alk. paper)
  1. Pragmatics.   2. Intention (Logic)   I. Kecskés, István.   II. Mey,
Jacob.
  P99.4.P72I575   2008
  401'.41−dc22
                                                              2008025685

Bibliographic information published by the Deutsche Nationalbibliothek

The Deutsche Nationalbibliothek lists this publication in the Deutsche Nationalbibliografie;
detailed bibliographic data are available in the Internet at http://dnb.d-nb.de.

ISBN 978-3-11-020606-7

Contents

Introduction

Istvan Kecskes and Jacob L. Mey

The Mouton Series in Pragmatics is committed to publishing books that attempt to explore new perspectives in pragmatics research. The present volume focuses on intention, common ground, and speaker-hearer behavior. These issues have been investigated by researchers for several decades; in recent years, however, some innovative approaches have been proposed that have shed new light on old issues, and the papers collected in this volume represent these new perspectives. It is important to note that the authors do not wish to reject existing views and theories. Rather, they attempt to revise them by adding new information and amalgamate old and new into a synergic whole.

In this introduction, we will not summarize the content of each paper as it is usually done in volumes such as this. Instead, we will briefly explicate the issues that the studies address, and give a short account of the status quo.

1. Intention

Recent studies (e.g., Verschueren 1999; Gibbs 2001; Arundale 2008; Haugh 2008) have pointed out that the role intention plays in communication may be more complex than proponents of current pragmatic theories have claimed. In particular, there is substantial recent evidence that militates against the continued placement of Gricean intentions at the center of pragmatic theories. While this evidence mainly comes from the socio-cultural, interactional line of research in pragmatics, the cognitive-philosophical line (such as represented by neo-Gricean Pragmatics, Relevance Theory, and Speech Act Theory) still maintains the centrality of intentions in communication. According to this view, communication is constituted by recipient design and intention recognition. The speaker's knowledge involves constructing a model of the hearer's knowledge relevant to the given situational context; conversely, the hearer's knowledge includes constructing a model of the speaker's knowledge relevant to the given situational context. Communication is supposed to be smooth if the speaker's intentions are recognized by the hearer through pragmatic inferences. Consequently, the main task of pragmatics is to explain how exactly the hearer makes these inferences, and determine what is considered the speaker's meaning. In a recent study, Levinson (2006) confirms that (Gricean) intention lies at the heart of communication, and proposes an "interaction engine" that underlines human interaction.

In contrast, the sociocultural-interactional paradigm considers intention to be 'problematic', and underlines its equivocality. According to this view, communication is not always dependent on speaker intentions in the Gricean sense (e.g., Verschueren 1999; Nuyts 2000; Mey 2001; Haugh 2008). In fact, one of the main differences between the cognitive-philosophical approach and the socio-cultural interactional approach is that the former considers intention an *a priori* mental state of speakers that underpins communication, while the latter regards intention as a *post factum* construct that is achieved jointly through the dynamic emergence of meaning in conversation. Since the two approaches represent two different perspectives, it would be difficult to reject either of them *in toto*. The complexity of the issue requires that we consider both the encoded and co-constructed sides of intention when analyzing the communicative process. In his paper in the present collection, Haugh proposes that the notion of intention need only be invoked in particular instances where it emerges as a *post factum* construct, salient to the interactional achievement of implicatures.

2.   Common ground and egocentrism

Current pragmatic theories have failed to describe common ground in its complexity: they usually retain a 'communication-as-transfer-between-minds' view of language, and disregard the fact that disagreement and egocentrism of speaker-hearers are just as fundamental in communication as are agreement and cooperation (cf. Kecskes 2008).

Some researchers (e.g., Stalnaker 1978; Clark and Brennan 1991; Clark 1996) consider common ground a category of specialized mental representations that exist in the mind *a priori* to the actual communication process. As Arnseth and Solheim (2002) have pointed out, both Clark and Brennan's joint action model (1991) and Clark's contribution theory (1996) retain a communication-as-transfer-between-minds view of language, and treat intentions and goals as pre-existing psychological entities that are subsequently somehow formulated in language. In these theories, common ground is considered to be relatively static, as *a priori* mental representations of the interlocutors, on the basis of which they conduct successful communication in a cooperative way, while their intentions are warranted.

The other approach to common ground has emerged as a result of recent research in cognitive psychology, linguistic pragmatics, and intercultural communication. Investigating how the mind works in the process of communication, cognitive researchers such as Barr and Keysar (2005), Colston (2005) have argued that mutual knowledge is not as significant as assumed by most people; instead, they formulated an emergence-through-use view of common ground, conceptualizing it as an emergent property of ordinary memory processes (e.g., Arnseth and Solheim 2002; Koschmann and LeBaron 2003). This dynamism is also emphasized in other studies (e.g. Heritage 1984; Arundale 1999), where it is stressed that real everyday communication is not dependent on recipient design and intention recognition, as it is claimed by most theories that have grown out of Grice's approach. The process is more like a trial-and-error, try-and-try-again-process that is co-constructed by the participants. It appears to be a non-summative and emergent interactional achievement (Arundale 1999, 2008).

With this dynamic revision of common ground, the role of cooperation is also challenged. Investigating intercultural communication, Kecskes (2007) argues that in the first phase of the communicative process, instead of looking for common ground, lingua franca speakers articulate their own thoughts with the linguistic means that are easily available to them. Earlier, Barr and Keysar (2005) had claimed that speakers and listeners commonly violate their mutual knowledge when they produce and understand language. Their behavior is called 'egocentric', because it is rooted in the speakers' or listeners' own knowledge instead of in mutual knowledge. Other cognitive psychologists too (e.g., Keysar and Bly 1995; Giora 2003; Keysar 2007), have shown that speakers and listeners are egocentric to a surprising degree, and that individual, egocentric endeavors of interlocutors play a much more decisive role in the initial stages of production and comprehension than is envisioned by current pragmatic theories envision. This egocentric behavior is rooted in speakers' and listeners' relying more on their own knowledge than on mutual knowledge. People turn out to be poor estimators of what others know. Speakers usually underestimate the ambiguity and overestimate the effectiveness of their utterances (Keysar and Henly 2002).

Findings about the egocentric approach of interlocutors to communication are also confirmed by Giora's (1997, 2003) graded salience hypothesis and Kecskes' (2003, 2008) dynamic model of meaning. Interlocutors seem to consider their conversational experience more important than prevailing norms of informativeness. Giora's (2003) main argument is that knowledge of salient meanings plays a primary role in the process of using and comprehending language. She claims that "privileged meanings, meanings foremost on our mind, affect comprehension and production primarily, regardless of context or literality" (Giora 2003: 103). Kecskes' dynamic model of meaning (2008) similarly emphasizes that what the speaker says relies on prior conversational experience, as reflected in lexical choices in production; conversely, how the listener understands what is said in the actual situational context depends on his/her prior conversational experience with the lexical items used in the speaker's utterances. Smooth communication depends primarily on the match between the two. Cooperation, relevance, and reliance on possible mutual knowledge come into play only after the speaker's ego is satisfied and the listener's egocentric, most salient interpretation is processed. Barr and Keysar (2005) argue that mutual knowledge is most likely implemented as a mechanism for detecting and correcting errors, rather than as an intrinsic, routine process of the language processor.

Papers in the second part of the volume explore and discuss both these sides of common ground. Here, it is important to emphasize that neither the ideal interaction approach of Clark and his followers nor the cognitive psychologists' approach appear wholly convincing when taken by themselves. Common ground comprises both *a priori* and *post factum* elements. Consequently, the egocentrism of interlocutors may be dominating in certain phases of the communicative process (where reliance on a priori elements happens to be more important) than it is in other phases of the same communicative process. While the papers in this volume represent a move in the right direction, further research is needed to investigate this complex issue.

References

Arnseth, Hans Christian and Ivar Solheim
    2002.   Making sense of shared knowledge In *Proceedings of CSCL 2002, Computer Support for Collaborative Learning: Foundations for a CSCL Community*, G.Stahl (ed.), 102-110. Boulder, CO. January 2002.

Arundale, Robert B.
    1999   An alternative model and ideology of communication for an alternative to politeness theory. *Pragmatics* 9: 119-154.
    2008.   Against (Gricean) intentions at the heart of human interaction. *Intercultural Pragmatics* 5(2): 231-256.

Barr, Dale J., and Boaz Keysar
    2005   Making sense of how we make sense: The Paradox of egocentrism in language use. In *Figurative language comprehension*, Herbert, L. Colston and Albert N. Katz. (eds.), 21-43. Mahwah, NJ: Lawrence Erlbaum.

Clark, Herbert H.
    1996   *Using Language*. Cambridge: Cambridge University Press.

Clark, Herbert H., and Susan E. Brennan
    1991   Grounding in Communication. In *Perspectives on socially shared cognition*, Lauren B. Resnick, Jonathan M. Levine, and Stephanie Teasley, (eds.), 127-149.Washington: American Psychological Association.

Colston, Herbert L.
    2005   On sociocultural and nonliteral: A synopsis and a prophesy. In *Figurative language comprehension: Social and cultural influences*, Herbert Colston and Albert N. Katz (eds.), 1-20. Mahwah, NJ: Lawrence Erlbaum.

Gibbs, Raymond W. Jr.
    1999   *Intentions in the Experience of Meaning*. Cambridge: Cambridge University Press.
    2001   Intentions as emergent products of social interactions. In *Intentions and Intentionality*, Bertram F. Malle, Louis J. Moses and Dare J. Baldwin (eds.), 105-122. Cambridge, MA: MIT Press.

Giora, Rachel
    1997   Understanding figurative and literal language. The graded salience hypothesis. *Cognitive Linguistics* 7: 183-206.

2003   *On Our Mind: Salience, Context and Figurative Language.* Oxford: Oxford University Press.

Grice, H. Paul
1975   Logic and conversation. In *Syntax and Semantics.* Volume 3: *Speech Acts*, Peter Cole and Jerry Morgan (eds.), 41-58. New York: Academic Press

Haugh, Michael
2008   Intention and diverging interpretings of implicature in the "uncovered meat" sermon. *Intercultural Pragmatics.* 5(2): 201-230.

Heritage, John
1984   *Garfinkel and Ethnomethodology.* Cambridge: Polity Press.
1990/91   Intention, meaning and strategy: observations on constraints from conversation analysis. *Research on Language and Social Interaction* 24: 311-332.

Kecskes, Istvan
2003   *Situation-Bound Utterances in L1 and L2.* Berlin/New York: Mouton.
2007   Formulaic Language in English Lingua Franca. In *Explorations in Pragmatics*, Istvan Kecskes and Laurence Horn (eds.), 191-218. Berlin/New York: Mouton de Gruyter.
2008   Dueling context: A dynamic model of meaning. *Journal of Pragmatics* 40 (3): 385-406.

Keysar, Boaz
2007   Communication and miscommunication: The role of egocentric processes. *Intercultural Pragmatics* 4 (1): 71-84.

Keysar, Boaz, and Bridget Bly
1995   Intuitions of the transparency of idioms: Can one keep a secret by spilling the beans? *Journal of Memory and Language* 34(1): 89-109.

Keysar, Boaz and Anne S. Henly
2002   Speakers' overestimation of their effectiveness. *Psychological Science* 13: 207-212.

Koschmann, Timothy and Curtis D. LeBaron
2003   Reconsidering Common Ground: Examining Clark's Contribution Theory in the OR. In *Proceedings of the Eight European Conference on Computer-Supported Cooperative Work*, Kari Kuutti, Eija Helena Karsten, Geraldine Fitzpatrick, Paul Dourish and Kjeld Schmidt (eds.), 81-98. Dordrecht/Botson/ London: Kluwer.

Levinson, Stephen C.
2006   Cognition at the heart of human interaction. *Discourse Studies* 8: 85-93.

Mey, Jacob L.
   2001   *Pragmatics. An Introduction* 2d ed. Oxford/Malden, MA: Blackwell.
Nuyts, Jan
   2000   Intentionality. In *Handbook of Pragmatics*, Jef Verschueren, Jan-Ola
             Östman, Jan Blommaert and Chris Bulcaen (eds.), Amsterdam: John
             Benjamins.
Stalnaker, Robert
   1978   Assertion. In *Syntax and Semantics.* Volume 9: *Pragmatics*, Peter
             Cole (ed.), 315-332. New York: Academic Press.
Verschueren, Jef
   1999   *Understanding Pragmatics*. London: Hodder Arnold.

Psychological explanations in Gricean pragmatics and Frege's legacy[1]

Katarzyna M. Jaszczolt

1. Introduction

At the end of the 19th century, Gottlob Frege developed a new concept of logic. In his *Begriffsschrift* (Frege 1879), he presented a function/argument analysis of the logical form of judgments according to which the reference of a predicate is a function from objects to truth values. This proposal marked the beginning of modern logic and breaking away from psychological logic, which focused on the study of thought processes and subjective mental representations. Frege's writings abound in arguments against psychologism in logic. While his task to ban psychologism from logic can be considered accomplished, at least for mainstream modern conception of it[2], there is a closely related question that remains unanswered: since Frege's logic provides the theoretical foundations for the conception of meaning adopted in truth-conditional semantics, does Frege's ban on psychologism extend to the theory of natural language meaning? In other words, does it extend to truth-conditional semantics and to its contextualist Gricean extension that is often called truth-conditional pragmatics?

In this paper I shall concentrate (apart from the final section of semantic minimalism) on the contextualist, post-Gricean approach to meaning according to which the result of pragmatic inference can contribute to the truth-conditional content as a free, top-down process of modulating the content – a process that is not restrained by the syntactic form of the expression (see, e.g., Recanati 2004, 2005). I identify four areas in which moderate psychologism is necessary in order to obtain a pragmatic theory that can be subjected to experimental testing:

[1] The selection of the perspective to be adopted: that of the speaker, the addressee, or a Model Speaker – Model Addressee interaction;

[2] The unit on which the pragmatic inference or default enrichment operates;

[3]   The definition and delimitation of automatic (default) interpretations vis-à-vis conscious pragmatic inference;

[4]   The definition of the object of analysis (called here Primary Meaning).

For the purpose of this discussion, I analyze two post-Gricean approaches to meaning: Levinson's (2000) theory of presumptive meanings and my Default Semantics (Jaszczolt 2005) and demonstrate that they both fulfill [1] with the help of considerations from processing; the theory of presumptive meanings is deficient in [2] and [4] and lacking in [3], and Default Semantics is at present deficient in [2] and lacking in [3] through shunning psychological explanations.

The structure of the paper is as follows. In Section 2, I present Frege's arguments against psychological claims in logic that will constitute the background for the discussion of psychologism in post-Gricean pragmatics and point out that, pace some attempts to do so, the transfer of this ban to pragmatics is by no means automatic. In Sections 3–6 I address the need for psychological explanations in the four areas identified above. I further justify the need for arguments from utterance processing by pointing out, in Section 6, that moderate psychologism in pragmatics is perfectly compatible with Frege's exclusion of psychology from logic and his arguments against psychological logic. In Section 7 I contend that even minimalist semantic accounts, as long as they are proposition-based, cannot escape psychologism.

In post-Gricean pragmatics, one has a choice of adopting one of the two perspectives on speaker's intentions: either (a) remain close to Grice and neo-Griceans and assume that pragmatic theory should offer a model of utterance interpretation that accounts for the meaning that can be plausibly construed as that intended by a speaker (Grice 1975), or (b) assume that pragmatic theory should model intentions as they are recovered by the addressee – a view represented by relevance theory (Sperber and Wilson [1986]1995). In the last two decades relevance theorists have argued for the advantages of the latter construal of pragmatics over the first. However, as Saul (2002) suggests, these two projects allegedly have fundamentally different objectives and do not yield to a relative evaluation. On her reading of Grice, Grice's notion of *what is said* makes it possible that both the speaker and the addressee are wrong about what is said. Grice's what is said (his meaning$_{nn}$) closely depends on the sentence and on the possible enrichments of the sentence that can be accounted for by rules capturing the regularities in rational conversational interaction. It is speaker's meaning but not in the sense of mental representations and processing of intentions. Instead, it is speaker's meaning in the sense of a typical, model speaker and the meanings this speaker conveys to the model addressee. On the other hand, relevance theorists seek an account of utterance processing: an account of the psychology of utterance interpretation. According to Saul, these projects are very different and can happily coexist.

In this paper I support a perspective on utterance meaning that is compatible with Grice's position and present it as a model of utterance interpretation on which an utterance by a *Model Speaker* is recovered by a *Model Addressee*. But unlike Saul, I argue that adopting Grice's position cannot mean shedding the interest in the psychology of processing. I focus in this paper on the so-called cultural defaults: cultural assumptions that constitute shortcuts through pragmatic inference. I demonstrate in the example of cultural defaults in utterance interpretation that neither of the two views can entirely avoid psychologism in pragmatics. The *process* of utterance interpretation figures as an essential explanatory component in the intention-based post-Gricean pragmatics if we want to provide a satisfactory account of utterance comprehension, including an account of pragmatic inference and default interpretations. Such defaults can, in principle, be understood in three different ways: as (i) defaults intended by the speaker and recovered by the addressee; (ii) intended by the speaker and not recovered by the addressee; and (iii) not intended but recovered. In order to select the perspective that is most adequate from the methodological point of view, one needs to resort to the discussion of utterance processing. I will argue that one cannot provide a pragmatic theory with an adequate account of utterance meaning without stepping down, at least in this introductory stage, to the issues of processing.

2. Psycholigism: a "corrupting intrusion"?

In his *Grundlagen der Arithmetik*, Frege (1884) takes a firm stance against psychological explanations in logic, which gave rise to entirely new understanding of truth, logic, definitions, objects of study of logic and mathematics, and also shed some light on linguistic theory. Frege (1884a: 90) says that "[t]here must be a sharp separation of the psychological from the logical, the subjective from the objective." He distinguishes concepts and objects on one hand, and ideas ('Vorstellungen') on the other, where the latter are psychological constructs and fall outside objectual investigations proper. Ideas are not to be confused with objects. The way people represent an object to themselves is not to be confused with the object itself. For example,

> If someone feels obliged to give a definition, and yet cannot do so, then he will at least describe the way in which the object or concept is arrived at. This case is easily recognized by the absence of any further mention of such an explanation. For teaching purposes, such an introduction to things is quite in order; only it should always be clearly distinguished from a definition. (Frege 1884a: 89).

In logic this means that checking the correctness of a definition or validity of a proof must not make use of psychological notions such as "thinking that something is true or valid" but instead one must resort directly to the laws of logic. At the time this was a landmark in philosophy, logic and arithmetic: a successful rebuttal of Husserl's view that logic is concerned with mental processes.[3] Frege (1884a: 88) writes:

> The description of the origin of an idea should not be taken for a definition, nor should the account of the mental and physical conditions for becoming aware of a proposition be taken for a proof, and nor should the discovery ['Gedachtwerden'] of a proposition be confused with its truth! We must be reminded, it seems, that a proposition just as little ceases to be true when I am no longer thinking of it as the Sun is extinguished when I close my eyes.

Frege's attack on psychologism permeates many of his works. In his *Grundgesetze der Arithmetik*, vol. 1 (Frege 1893), he dubs the effect of psychology on logic a "corrupting intrusion" (202). He repeats that "*being true* is quite different from *being held as true*" (202). Similarly, in *Logic*, Frege ([1897]1969) presents the task of logic as "isolating what is logical ... so that we should consciously distinguish the logical from what is attached to it in the way of ideas and feelings" (243), because "Logic is concerned with the laws of truth, not with the laws of holding something to be true, not with the question of how people think, but with the question of how they must think if they are not to miss the truth." ([1897]1969:250).

Frege's argument hardly requires further defense. His aim was successfully accomplished.[4] Modern logic dominated old phenomenological approaches and the function/argument analysis of judgments[5] established firm foundations for formal analyses of artificial and subsequently natural languages, developed by Tarski and Davidson respectively. The parallel emergence of structuralist linguistics took it one step further. Just as the history of thinking about an object fell outside the science proper, so de Saussure's synchronic/diachronic distinction established firm foundations for the dissociation of the history of a phenomenon from theorizing about it as an object in itself.

Truth-conditional semantics became established in the dominant orientation for analyzing natural language meaning. But when the boundaries of truth-conditional semantics became blurred by Gricean and post-Gricean attempts to make various types and degrees of pragmatic inference "intrude" into the unit of which truth conditions should be legitimately predicated[6], the question of psychologism resurfaced. The applicability of Frege's arguments to post-Gricean analyses of meaning is not direct or straightforward. While psychologism is to be banned from definitions in logic, it need not always be banned from explanations in a subject matter that uses logic as a tool. Our task is to assess whether, and if so, to what extent Frege's arguments are applicable to the current theories of meaning, and to post-Gricean pragmatics of a propositional, truth-conditional variety in particular. The kinds of questions that arise are:

> Q1: Should psychological explanations be present in definitions of what is said, truth-conditional content, the explicit/implicit distinction, etc.?

> Q2: Should the psychology of utterance processing be considered in the discussions of the boundary between semantics and pragmatics?

In the course of this paper I shall give a positive answer to these and related questions by means of investigating the identified areas [1] – [4] one by one. I will conclude that a moderate dose of psychologism is a necessary feature of any Gricean, truth-conditions-based, theory of meaning.

3.   Whose perspective?

I move now to the discussion of the identified area [1]: the perspective to be adopted in modeling utterance meaning. Frege writes about Husserl's psychological treatment of number in his review of Husserl's *Philosophy of Arithmetic, I:*

> If a geographer was given an oceanographic treatise to read which gave a psychological explanation of the origin of the oceans, he would undoubtedly get the impression that the author had missed the mark and shot past the thing itself in a most peculiar way. (…) The ocean is of course something real and a number is not; but this does not prevent a number from being something objective, and this is what is important. Reading this

> work has enabled me to gauge the extent of the devastation caused by the
> irruption of psychology into logic… (Frege 1894: 209).

It is difficult to disagree. But it does not follow that psychology causes an equal "devastation" to pragmatics. What I mean by this caveat is this. The definitions of the proposition, utterance meaning, default enrichment, pragmatic inference, must be established first and this is to be done with the help of the distinction between the three perspectives: that of the speaker, the addressee, and Model Speaker – Model Addressee. In other words, one has to decide whose meaning one wants to model. This is particularly important when the interlocutors differ with respect to the background cultural knowledge and the common ground is wrongly assessed.[7] At this stage psychology plays an essential part, and, indeed, the psychology of utterance processing is the main object of the argument. Constructing the object of study involves these psychological considerations necessarily. It is only at the subsequent stage, once the orientation is selected, that we can proceed either way: stay close to Frege and opt for (I) a theory of a Model Speaker – Model Addressee interaction, or depart in the more radically psychological direction for the remaining two positions: theories of (II) speaker's or (III) addressee's mental representations. The decision concerning the perspective is not, however, a decision as to *whether* to admit psychologism into pragmatic theory. Rather, it is a decision *within* the domain of the psychology of processing. Perspectives (I)–(III) are all contaminated, so to speak, albeit to different degrees.

It should also be pointed out that option (I) is the closest to Frege in spirit in that Fregean thoughts are not psychologically real ideas of a person but rather entities of a different realm, external to particular minds and brains. Also, Frege's analysis of judgments into function and argument calls for a notion of a function from objects to judgments (true or false) which is independent of the actual judgments made by judging agents. The analysis has to account for assertions that have never been made, allowing for formalizing those that could have been made, those yet to be made, etc.[8] Green (2003: 224), for example, claims that "the notion of a function is derivative of the prior grasp of what it is to be sensitive to the presence of a property." However, it is humankind's "sensitivity to properties" that makes up Frege's concepts, rather than the particular sensitivities of particular minds.

The final point on the choice among (I)–(III) concerns their compatibility. One may be tempted to extend Saul's (2002) claim,[9] that Gricean program and relevance theory can happily co-exist, to all three of the above standpoints and say that all three can happily co-exist: they have different, compatible objectives. I will refer to this thesis as the Co-existence Thesis. However, as I have argued here, this co-existence would be possible if not for the fact that all three positions involve some degree of psychologism. We are not faced with a choice between (a) two possible types of accounts of utterance processing, speaker- and addressee- oriented, vs. (b) Grice's account of meaning$_{nn}$, free from considerations of processing. If we were, the co-existence of (I) with (II) or (I) with (III) would be defensible. Grice's very distinction between context-free and context-dependent implicatures, as well as his construal of implicature as a post-propositional *process* (in his original sense of implicature) of recovering an *implicatum* testify to the interest in the psychology of the recovery and construction of utterance meaning: an interest in the questions as to whether it is automatic or not (my identified area [3]), and at what stage in processing the pragmatic meaning is added (my identified area [2]). And once we take one step in the direction of psychologism, we have to take another, in the interest of consistency and psychological accuracy. Once one postulates automatic, presumed, default meanings, one needs an empirically falsifiable account of what exactly counts as automatic and default.[10] Analogously, once one commits oneself to local (cf. Levinson) or global post-propositional (cf. Grice) defaults, one needs an empirically falsifiable account of how local or how global they are to be. It is for this reason that in Default Semantics the primary meanings of utterances, and their formal equivalents in merger representations, can be more global than a proposition itself and be built out of larger units of discourse – à la discourse representation structures (DRSs) in Discourse Representation Theory (DRT, Kamp and Reyle 1993; van Eijck and Kamp 1997; Kamp *et al.* forthcoming). But how local pragmatic operations can be remains unanswered at present. Resorting to hypotheses about processing is sometimes the only way to secure the object for experimental testing.

I attend to these big questions in Sections 4 and 5. For the remainder of this section I shall move to proposing and tentatively defending option (I), the Model Speaker – Model Addressee meaning, as the most promising perspective.

Grice's *what is said*, as Bach (1994) and Saul (2002) well explain, is not very well suited as a technical term for a cognitive theory of utterance processing. When the speaker's intentions are incompatible with the conventional meaning of the sentence, the speaker's meaning becomes dissociated from what is said: the speaker makes as if to say what his or her utterance of the sentence conventionally conveys. Similarly, the addressee's meaning can become dissociated from what is said: the addressee may misunderstand the speaker's intentions. What we are left with is a theoretical notion of what is said that is supposed to fulfill two tasks: (1) capture the fact that speakers are not free to convey any content by any randomly chosen utterance but rather are constrained by the conventions of sentence-type meaning and (2) provide a unit of meaning that combines sentence meaning with those utterance-specific embellishments that his theory allows (i.e., disambiguation and reference assignment; see Grice 1978).

This theoretical construct of what is said can happily exist without making recourses to the psychology of utterance processing. However, as we know from the last thirty years of pragmatics research, this construct is faulty in many respects. First, it is an arbitrary decision to draw a line between reference assignment to indexical expressions and disambiguation on the one hand, and the wide array of other pragmatic additions to the meaning of the sentence. Next, tying the notion of saying to sentence's conventional meaning is at odds with any concept of utterance meaning, be it speaker's, addressee's, or neither. Instead, one can entertain the following construal. The theory of utterance meaning (meaning$_{nn}$) is to account for the meanings the interlocutors *normally* convey and at the same time the meanings the addressees *normally* recover in the process of a rational conversational interaction–a cooperative interaction that is founded on principles (maxims, heuristics) that capture the regularities of their behavior, where the latter is assumed to be rational. On this construal, we have a Model Speaker and a Model Addressee and proceed to explain *why* it is so easy to communicate one's thoughts. What we don't capture is the cases when it is *not* easy: the cases when there is a mismatch between the intended and the recovered meaning. They can occur when, for example, a cultural assumption is incorrectly taken by the speaker to be shared by the interlocutors, when an implicature is intended but not recovered, or recovered when not intended, such as in the cases of misconstrued sarcasm or irony. In these cases there are misconceptions about the common ground, where by common ground I will understand the set of cultural, social, and other assumptions taken by the interlocutors to be shared, as well as the particular context of conversation that gives rise to shared information.[11]

This omission is a justified one. We abstain from making unsupported hypotheses about processing, the theory is free from contamination from psychologism and we have a generally adequate account of how speakers *normally* externalize their intentions, allowing even for the context-dependent particularized implicatures which are catered for as inferences from the proposition meant and which, as I argue in *Default Semantics* (Jaszczolt 2005a) and Section 6 below, can function as the primary intended content. They are regarded as primary meanings on the grounds of their salience, irrespective of the criteria laid out in the distinctions between said and implied or explicit and implicit. We can even cater for the constraints in (1) and (2) above in that there is nothing to stop us from increasing the range of permissible enrichments à la Carston (1988) and relaxing the concept of saying à la Bach (1994) in search for an adequate unit on which pragmatic inference and defaults are founded. The search for such a unit is identified as the second area ([2] above) in which psychological considerations have a role to play and is discussed in the following section.

For the purpose of the discussion of the three identified areas for psychology of processing, I use the example of conversational defaults in Levinson's (2000) and Jaszczolt's (2005) theories. Cooperation in conversation is based on the manifestation of speaker's intentions which are recovered by the addressee either by means of pragmatic inference or by relying on shortcuts through such a process, called among others default interpretations (Asher and Lascarides 2003, Jaszczolt 2005a) or presumptive meanings (Levinson 2000). Default interpretations, understood as a theory-independent, common-sense category of salient, unmarked meanings, are hardly controversial[12]. *A fortiori*, by default reasoning (no pun intended), my argument amounts to the claim that any Gricean theory of utterance meaning must include a dose of psychologism and attend to the differences between (a) conscious, costly, inferential and (b) automatic processes.

Let us consider first Levinson's (1995, 2000) presumptive meanings, utterance-type meanings, or, as he calls them after Grice, generalized conversational implicatures (GCIs). As I explain in Section 4, Levinson's GCIs are not identical with Grice's original conception but this point will not have to concern us at this moment. What is important is that for both authors GCIs belong with Model Speaker's meaning. They are not default inferences performed by the addressee and hence do not belong with addressee's meaning.[13] Neither do they belong with the speaker in the sense of being part of the mental representation or any other psychologically-loaded phenomenon. They belong with speaker's meaning in a processing-free sense of a theory of linguistic competence where we can talk about normal, *ceteris paribus* inferences. For Levinson (2000: 1), "Utterance-type meanings are matters of preferred interpretations (…) which are carried by the structure of utterances, given the structure of the language, and not by virtue of the particular contexts of utterance."

They constitute a middle level between sentence meaning and utterance-token meaning. They are triggered by a set of three heuristics – a set of generalizations over rational conversational behavior which amounts to an improved set of Grice's maxims of conversation. But in order to have such Model Speaker – Model Addressee theory of defaults, one has to be in a position to say (a) what unit triggers such defaults and (b) what exactly counts as default, as opposed to inferentially achieved, interpretation. We cannot provide this answer by direct experimentation alone. In order to design experiments, one needs intuitively plausible hypotheses, theories to be tested. Such a discrimination between extant theories, or, even better, constructing more plausible ones, cannot proceed without answering the question "what would processing have to be like for this claim to be true?" Only then do we provide food for experimentation. In other words, we also have to have answers in areas [2] and [3].

Let us move now to Default Semantics. Like Levinson's account, it adopts the Model Speaker – Model Addressee orientation. So, the discussion of the former applies to the latter as far as our identified area [1] is concerned.

To conclude, I proposed in this section that while Gricean theory of meaning$_{nn}$ respects Frege's rejection of psychological explanations from logic, one must resort to psychology in choosing and defining this perspective and thereby choosing and defining the object of study. I have also pointed out that the choice of the object of study is not a free choice founded on individual interests. I have argued against Saul's claim, and against an extended Saul's claim which I called the Co-Existence Thesis, that the three possibilities of construing a theory of utterance meaning can co-exist. An important task for the Gricean pragmaticist is to choose that option from among (I)–(III) that is methodologically superior by the well-established criteria of adequacy of a scientific theory. I argued that option (I), a theory of meaning for a Model Speaker – Model Addressee interaction, is such a methodologically preferred position. I shall finish this section on a Fregean conciliatory note. Just as Frege, in *Der Gedanke* (1918–1919), writes that the task of logic is not "investigating minds and contents of consciousness owned by individual men;" the task is "the investigation of *the* mind; of *the* mind, not of minds" (342), so a theory of meaning$_{nn}$ is the investigation of *the* utterance meaning; *the* meaning, not meanings for speakers or addressees. Similarly, just as for Frege (1918–1919: 342) "[a]lthough the thought does not belong with the contents of the thinker's consciousness, there must be something in his consciousness that is aimed at the thought," so, for a post-Gricean, although utterance meaning does not belong with the mental processing of an utterance by the interlocutors, there must be something in mental processes that is aimed at the utterance meaning. This something, Frege warns, should not be confused with the object of study. But post-Gricean pragmatics need not be at fault in this respect and indeed neither of the two accounts discussed above (Levinson 2000 and Jaszczolt 2005a) is culpable of such confusion. As I have argued here, the opposite is the case: we need more psychological considerations in choosing the perspective from which meaning should be modeled.   Only when this is achieved and experimentally tested may Frege's arguments for the separation from psychology begin to apply.

4.   Pragmatic increments

In this section I address the identified area [2]: the unit on which pragmatic inference or automatic enrichment operates in utterance interpretation and by the same token the unit that should be adopted as a basis for such pragmatic modifications in pragmatic theory.

A modest recourse to utterance processing is a condition *sine qua non* for any Gricean theory of utterance meaning when it tries to answer the question as to at what stage in the interpretation of discourse inferential and automatic enrichments take place. Questioning the recourse to processing by questioning the need for pragmatic theory to answer this question would have direct consequences on the Gricean program of providing a theory of utterance meaning, and hence constitutes too radical a solution.

Let us consider Levinson's presumptive meanings in his theory of generalized conversational implicature. Grice's GCIs are based on the proposition said (in his restricted sense of *what is said*[14]). For Levinson, GCIs can arise locally, pre-propositionally: sentence meaning does not have to be *processed first*. This assumption of localism is an essential step in introducing and defining the concept of GCIs/presumptive meanings. But does localism adhere to the processing-free conception of pragmatic theory? One can levy the following objection at this point. While apparently abstaining from using processing as an *explanans*, Levinson has to include it as an *explanandum* in order to introduce his notion of presumptive meanings:

> Explicit processing considerations do not enter the framework offered here, but they do form part of the background, for the character of the inferences in question as default inferences can, I think, be understood best against the background of cognitive processing. The evidence, so far as it goes, from the psycholinguistic literature is that hypotheses about meaning are entertained incrementally – as the words come in, as it were. (Levinson 2000: 5).

What this means is that the enrichment to the default meaning takes place as soon as the relevant item is processed – be it a word, a phrase, a whole proposition, or, indeed, a morpheme: "…for example, a scalar quantifier like *some* will, as I will show, already invoke default enrichments before the predicate is available." (Levinson 2000: 5).

> Examples in (1)–(3) emphasize the importance of locality for Levinson's GCIs.

> (1) Some of the boys came. +> "not all"

> (2) Possibly, there's life on Mars. +> "not certainly".

> (3) If John comes, I'll go. +> "maybe he will, maybe he won't"

> (from Levinson 2000: 36–37).

These examples are all captured by the Q-heuristic, "What isn't said, isn't": if a stronger expression is available and was not used, infer that it is not the case. As the glosses of what is conversationally communicated (+>) indicate, the presumptive meaning is triggered by the smallest relevant item: "some" triggers "not all", rather than "Some of the boys came" triggering "Not all of the boys came" as it was the case on Grice's original construal.

Examples subsumed under the working of the I-heuristic ("What is expressed simply is stereotypically exemplified") make even more use of the feature of locality:

> (4) bread knife +> knife used for cutting bread

> kitchen knife +> knife used for preparing food, e.g. chopping

> steel knife +> knife made of steel

(5) a secretary   +> female one

(6) a road +> hard-surfaced one

(7) I don't like garlic. +> I dislike garlic. [triggered by "don't like", KJ]

(from Levinson 2000: 37–38, adapted)

There are two problems here. First, examples in (4) are compounds and therefore they are better regarded as lexical entries than examples of local default enrichment.[15] Second, examples (5) and (6) are contentious. For example, "the Prime Minister's secretary" does not seem to trigger an enrichment to "female" and hence the default status of this interpretation is doubtful. "Road" is a lexical item that comes with the conceptual baggage of a prototype, definition, set of features, and so forth, depending on one's favored approach to word meaning, and does not seem to undergo local enrichment. *Neg*-raising in (7), on the other hand, is a wide-spread fact of conversation but it is not obviously local: the shift in the interpretation to "dislike" seems to be triggered by the content, the proposition. One can rebut these objections by pointing out that Levinson's theory of GCIs is a theory of utterance-type meaning and as such it can easily be dissociated from the observations on the incremental nature of utterance processing made by psycholinguists.[16] After all, its main preoccupation is with the modularity of semantics and pragmatics and with the differentiation of the middle level of GCIs from both modules. This level is not pragmatic because it does not involve computation of speaker's intentions, neither is it semantic because it does involve default inferences from the output of grammatical processing. GCIs are not Bach's (1995) standardized, routine meanings either: new meanings can also be captured by the three heuristics. They are, as a product of some or other process of utterance interpretation, propositional. It is this emphasis on the product that allows Levinson to eschew questions about processing. But does it allow him to do so successfully? Some presumptive meanings are amenable to being construed as local, with no claim about processing being included. Possible as it seems, this is not what Levinson's theory does. As the quotation from Levinson (2000: 5) clearly shows, Levinson *is* concerned with matters of processing. Cognitive adequacy of his presumptive meanings is an important goal.

The final problem concerns default cancellation, or defeasibility.[17] The more local the enrichments, the higher the likelihood that they have to be taken back later on in discourse, when more information becomes available. For example, the default reading of "some" has to be cancelled in (1'):

(1') Some (+> "not all") of the boys came. In fact, all of them did.

It has to be cancelled even sooner in (1''):

(1'') Some (+> "not all"), in fact all, of the boys came.

Cancellation is costly. If we remained close to Grice's original concept of an implicature and adopted post-propositional, global enrichments, the cost and frequency of such cancellations would be substantially reduced but would not disappear completely. If we went further and allowed implicatures (including the default ones) to be founded on units even larger than a sentence (proposition) when this is appropriate, i.e. when there is clear evidence that this is so, then we would come even closer to an intuitively adequate model of utterance interpretation. What we ultimately want is a theory that construes pragmatic inference and defaults as operating on a unit that is adequate for this particular discourse. And by "adequate" I mean here cognitive, psychological adequacy, which can be expected to be corroborated by testing of the processing.

In Default Semantics, the problem of combining information from stereotypes with the meaning derived from the sentence is approached in the following way. When we assume a pragmatic approach to the compositionality of meaning (Recanati 2004, Jaszczolt 2005a), such cultural defaults need *not* pertain to the enrichment of the logical form understood as the output of syntactic processing but they can also override it. This view results in a notion of *what is said* which is more psychologically plausible from the point of view of accounting for intentions, and at the same time does not suffer from the problem of justification of the enrichment of the logical form as a different concept from implicatures (logically and functionally independent logical forms; Carston 1988, 2002). In composing utterance meaning, the output of syntactic processing is not pragmatically enriched but instead all the sources of information about meaning are equal contributors to the so-called *merger representation*.[18,19] The primary meaning of an utterance is understood as the meaning that is intended by a model speaker and recovered by a model addressee – a result of, and a contribution to, the assumed common ground. If the result of merging the sentence with the information from default interpretations is a representation that does not resemble the logical form of the uttered sentence, there is no special case there that would be in need of an explanation: the resemblance is not the norm, and neither is there a requirement that the primary meaning of the utterance has to be a proposition that entails the proposition uttered or be a development of the logical form of the sentence. Primary meaning is the most salient meaning that can be assumed to be intended and recovered as intended, period.

A further advantage of merger representations comes from their dynamic-semantic foundation. Just like DRSs of DRT, they can collect information incrementally across sentence boundaries. They are representations of discourses, not sentences. When a proposition that corresponds to the speaker's meaning relies on more than one sentence, this situation can easily be accommodated in a merger representation. Anaphoric dependencies carry on intersententially. The condition of flexibility of the unit to which defaults and inference pertain is fulfilled, and we can conclude that merger representations seem to constitute the food for experimentation that we were looking for. In Section 5 I argue that merger representations of Default Semantics (Jaszczolt 2005a, 2006b), loosely modeled on DRSs of DRT, fit the role of such flexible units in principle, although no general hypotheses concerning the length of the unit have been proposed to this point yet. The framework allows for local as well as global enrichment but the principles are not yet delineated for making generalizations concerning the length of the unit on which inference or automatic enrichment operate. Therefore, as a safe bet, all enrichment is considered to be global, post-propositional, until a more detailed hypothesis can be worked out and submitted for experimental testing.

To sum up at this point, hypotheses about processing are necessary in specifying the triggering unit for automatic enrichment. What is important for the current argument is the order in which theoretical explanations and experiments are to be placed. Before we can design reliable experiments, we need theories, hypotheses to be tested. We need a theory that would be more relaxed about the locality of enrichments than Levinson's presumptive meanings and more informative than Default Semantics that takes enrichments to be global. The hypothesis that is to be tested should allow for different lengths of the input on which inference-based or default interpretation are formed. By "length" I mean here the variability between very local, less local, post-propositional, or even multi-propositional input as it is conceived of in DRT and in Default Semantics. At the same time, the theory should preserve the intuition that the cancellation of such interpretations is costly and unwelcome. All these theoretical assumptions can only be properly discussed when we are not made to shun psychologism and theorizing about processing. Theorizing comes first and establishes hypotheses; experiments can follow. Avoiding theoretical discussions of the issues of processing led to the current situation in which experimental pragmatics is starved of hypotheses to test. Current experimentation is largely confined to falsifying Levinson's account with its rather inflexible notion of GCIs (see Noveck and Sperber 2004). Instead, it would be more profitable to test hypotheses that use a variable input to salient meanings.

In his 2006b, Atlas expresses a view against what he calls "armchair psychologizing" and in favor of "empirical psychology of sentence-processing." What I am arguing for here is that "psychologizing" has to appear before empirical studies. It is necessary in selecting and defining the object of study before empirical psychologists know *what* to test. And if psychological explanations have to, temporarily, figure in definitions, so be it.

5.   Cultural defaults and cultural inferences

The third of our identified areas is drawing the boundary between automatic, default interpretations and conscious pragmatic inference. As soon as we try to model any cases of default interpretations, i.e., shortcuts, through pragmatic enrichment of the sentence's content, we stumble across the problem of their definition and delimitation. This is the problem that permeates Grice's GCIs and also Levinson's theory of presumptive meanings. The reason for this is their shunning psychological explanations.

Let us consider the case of the possessive in the form of a genitive NP[20] as in (8) and Levinson's multiple-choice gloss of it:

(8) John's book is good.   +> the one he read, wrote, borrowed, as appropriate.

(from Levinson 2000: 37)

There are two problems with this gloss. First, it is difficult to see how genitive NPs count as a case of a default, presumed meaning. This problem is analogous to the one we encountered in examples (5) and (6) above but appears even more prominently here. A fair amount of assumed background is needed before the enrichment can take place, and it seems plausible to assume that this enrichment is not automatic but instead takes the form of conscious pragmatic inference. Secondly, even if they are to be understood as triggering presumed meanings, it is not clear exactly which unit triggers this enrichment: is it the NP, or the entire sentence, or the genitive marking on the noun itself. Compare (9) and (10):

(9) Chomsky's book is about grammar.

(10) John's book won the Booker Prize.

Arguably, assuming the standard common ground that the interlocutors are likely to adopt here, (9) defaults to "the one he wrote", and, arguably, it does so as soon as the word "Chomsky's" is processed. In (10), "John's book" is likely to default to "the one he wrote" when the entire VP has been processed. It seems that we cannot give a comprehensive account of the GCIs pertaining to possessive constructions *unless* we resort to "increments" in processing that are of a variable length, depending on a particular circumstance. Neither can we progress any further with this account without speculating on the boundary between utterance-type and utterance-token. Where does the common ground fit in? Is (11) the case of utterance-type because it is plausible to assume that the interlocutors identify Leonardo with Leonardo da Vinci? And, how reliable is this move? Can we also apply this reasoning to "Larry," meaning Larry Horn, in (12)? If not, where do we place the boundary?

(11) Leonardo's painting was stolen from Czartoryskis' Museum in Kraków.

(12) Larry's book is a thrilling account of negation.

Let us now move to presumptive meanings subsumed under the M-heuristic: "What's said in an abnormal way isn't normal" and examples in (13)–(14):

(13) It's not impossible that the plane will be late. +> rather less likely than if one had said "It's possible that…"

(14) Bill caused the car to stop. +> indirectly, not in the normal way, e.g., by use of the emergency brake.

(from Levinson 2000: 39)

By comparing the utterances in (13) and (14) with some more standard ways of communicating the same content one can obtain the GCIs as above. However, once again, the question arises as to where to draw the boundary between these alleged GCIs and inference from the context. A related question is how do we decide whether enrichments that are triggered by a word are cases of M-triggered GCIs or simply lexical coding, as in (15)? If Tom acquired a particularly expensive car, the verb "purchased" may not trigger additional implicatures. On the other hand, (16) may do so, as indicated.

(15) Tom purchased a car.

(16) Tom purchased a hamster.   +> Tom blew the event of acquiring a hamster out of proportion.

A similar problem applies to the compounds in (4) above ("bread knife", "kitchen knife", and "steel knife") in that they can easily be construed as cases of lexical meaning.[21] It is quite possible that in pursuit of localism Levinson may have gone too far and subsumed lexical meanings under pragmatic enrichment.

To conclude at this point, psychological explanation (*followed by* empirical evidence) of processing is necessary in (i) discriminating between cases of default and inference-based enrichment, and (ii) discriminating between default inference and lexical meaning.

Finally, let us consider the default – inference boundary problem in the theory of Default Semantics. In Default Semantics, a typology of default interpretations is suggested and it includes a category of so-called *social-cultural defaults*: interpretations that arise without conscious inference thanks to shared information about culture and society. But assumptions concerning such sharing of cultural or social knowledge can be mistaken. For example, the exchange in (17) achieves a humorous effect due to such a mismatch of intended and recovered meaning.

(17) A: So, is this your first film?

  B: No, it's my twenty second.

  A: Any favourites among the twenty two?

  B: Working with Leonardo.

  A: da Vinci?

  B: DiCaprio.

A: Of course. And is he your favourite Italian director?

(Richard Curtis, *Notting Hill,* 1999)

I shall concentrate here on shared cultural assumptions and will call them, for this purpose, cultural defaults. To repeat, cultural defaults can, in principle, be construed in three different ways: as (i) defaults intended by the speaker and recovered by the addressee; (ii) intended by the speaker and not recovered by the addressee; and (iii) not intended but "recovered". Attending to all three cases would require a significant dose of psychologism in semantic theory – far more than a linguistic theory can benefit from. A more satisfactory solution is to opt for a particular perspective, for example that of the speaker. Or we can opt for a model that is closer to Grice (and Levinson) in spirit and construe an account of utterance meaning on which we try to discern those standard interpretations that are normally shared between the interlocutors.[22] This option corresponds to the perspective of a Model Speaker – Model Addressee utterance interpretation defended in Section 2. In this way defaults, presumptive meanings, unmarked interpretations, generalized implicatures, and so forth, are closer in spirit to what we intuitively understand as defaults: interpretations that can be safely assumed to go through. However, one of our previously discussed problems persists on this account. Such defaults are highly dependent on the context of the particular discourse. For example, we can say that "working with Leonardo" triggers for many people the referential individuation as Leonardo DiCaprio in (17) above because of the previous co-text of talking about acting in films, as well as the general context of an interview with a movie star.[23] So, there is a problem here of how much of the previous discourse are we allowed to consider while still retaining the concept of default for a particular interpretation rather than calling it a case of pragmatic inference. We need reliable criteria for discerning such reliable, shared interpretations. In other words, we need to establish where, at what level of specificity, the default meaning ends. In Levinson's (2000) example (18), it is assumed that the speaker and the addressee belong to a society in which nannies are normally female and the enriched, default interpretation occurs automatically, unreflectively.

(18) We advertised for a new nanny.   +> a female nanny.

However, as I pointed out elsewhere, "...what the speaker uttered is that they had advertised for a nanny, of unspecified sex, age, social status, marital status, hair colour, skin colour, religion, sexual preferences, etc. How far do we want to go in postulating defaults? And, more importantly, what would the criterion for such a default representation content be?" (Jaszczolt 2005a: 55).

Do we also want to include enrichment from our cultural and social stereotypes of nannies in the content of the default interpretation? If so, do we use Mary Poppins, Maria from *The Sound of Music*, or perhaps even Nanny McPhee? This question remains unanswered and, it seems that the best way to answer is an empirical investigation. It is necessary to find out what the processing of such utterances really entails, but only when the plausible food for experimentation is established: one needs a reliable criterion for distinguishing defaults from results of pragmatic inference and it seems that the only available move is a recourse to processing à la Recanati (2004).

We can conclude that both the theory of presumptive meanings and Default Semantics are wanting in the criteria for delimitation of automatic, default interpretations and that these criteria cannot be postulated in a void; they need arguments from psychology. They don't need arguments from experimental psychology because experimental psychology needs precisely this food for experimentation that good hypotheses can provide. Poor hypotheses are a waste of experimenters' time, as the evidence form the premature testing of locality triggered by Levinson (2000) clearly indicates: premature because neither locality, nor the demarcation between defaults and inference, were sufficiently well worked out theoretically.

6.   Primary meaning and Fregean thoughts

I proceed now to the identified area [4]: the definition and delimitation of the object of study, commonly known as "what is said", "utterance meaning", "explicit meaning", and so forth. I have already committed myself in Section 3 to the perspective of Model Speaker – Model Addressee and will therefore assume it and confine the discussion to the question of the explicit/implicit boundary, that is whether the primary, explicit meaning which is subjected to the truth-conditional analysis in post-Gricean contextualist pragmatics has to obey the syntactic constraint and constitute the development of the logical form of the uttered sentence.

This question is discussed in depth and with empirical support in a separate paper (Sysoeva and Jaszczolt 2007). For the aim at hand I shall flag the problem and present some arguments in favor of rejecting the syntactic constraint.

Within the contextualist orientation in post-Gricean pragmatics there are several suggestions for what is to count as the primary unit of analysis. The most prominent candidates are Recanati's (1989) what is said, relevance-theoretic (Sperber and Wilson 1986) explicature, and Bach's (1994) impliciture. We can also add here the default-semantic (Jaszczolt 2005) meaning merger, also called primary meaning (Sysoeva and Jaszczolt 2007). All these concepts are contextualist concepts in that they subscribe to some form of pragmatic enrichment of the truth-conditional content. In one of the more radical versions of contextualism, this enrichment is called modulation and is stipulated to be present in every case of utterance interpretation: "Contextualism ascribes to modulation a form of necessity which makes it ineliminable. *Without contextual modulation, no proposition could be expressed…*"(Recanati 2005: 179–180); and "…there is no level of meaning which is both (i) propositional (truth-evaluable) and (ii) minimalist (that is, unaffected by top-down factors)" (Recanati 2004: 90).

For example, (19) is the uttered sentence, corresponding to a so-called minimal proposition, and (20) is its modulated equivalent, or what is said.

(19) Mary hasn't eaten.

(20) Mary hasn't eaten breakfast yet.

According to the view represented in Default Semantics, there is indeed such a top-down process of pragmatic inference that interacts with the aspects of meaning provided by the sentence. It also interacts with the aspects of meaning provided by social and cultural assumptions that are added automatically. But not all utterances make use of this pragmatic process of modulation. In this respect the view is not contextualist in Recanati's strong sense, although contextualist in the more general sense of allowing for a wide range of free enrichments to the truth-conditional content.

Default Semantics also says that the object of study of a truth-conditional theory of utterance meaning is the primary meaning construed as intended by the Model Speaker and recovered by the Model Addressee. This primary meaning need not obey the syntactic constraint. In other words, it need not be dependent on the syntactic representation of the uttered sentence. The latter claim distinguishes Default Semantics from other contextualist frameworks. To compare, "What is said results from fleshing out the meaning of the sentence (which is like a semantic 'skeleton') so as to make it propositional" (Recanati 2004: 6). "An assumption communicated by an utterance U is *explicit* if and only if it is a development of the logical form encoded by U" (Sperber and Wilson [1986] 1995: 182).

Both concepts, what is said and explicature, obey the syntactic constraint. Similarly, Bach's "middle level of meaning," called an impliciture, goes *beyond* what is said in the manner restricted by the syntactic constraint (Bach 1994, 2001, 2004, 2005): the syntactic form is the skeleton on which it is built.

To repeat, modulation is a so-called top-down process which is pragmatically rather than syntactically controlled. In other words, the additions to the logical form of the sentence are not controlled by the structure of the sentence; they need not be confined to filling in syntactic slots. At this point one can ask whether this modulation should not be construed as being even more free from syntactic constraints. In addition to not being dictated by slots in the logical form, it seems that there is no reason not to model it as being free from the requirement of being a development of the logical form altogether. In defining the main object of study of a theory of utterance meaning it seems only natural to begin with the question what is the main, primary meaning that can be identified as intended and communicated. Bearing in mind the decision we made concerning [1], the question is further narrowed down as: What is the main, primary meaning that can be identified as intended and communicated by the Model Speaker to the Model Addressee? There is no intra-theoretic or external epistemological reason for this meaning to resemble closely the meaning of the uttered sentence. For example, (19) is frequently used to communicate (21), (22), or a range of other contents as the main intended message.

(21) Mary is hungry.

(22) Mary wants to go for a meal.

Examples (21) and (22) would normally be classified as strong implicatures of (19). But when (21) or (22) is the main intended meaning (primary meaning), it seems that this is the meaning that should constitute the main object of analysis in pragmatic theory, and by extension it should be modeled in contextualist truth-conditional semantics, which we also referred to as truth-conditional pragmatics.[24]

Default Semantics does not recognize a need for a syntactic constraint. It rests on the assumption that the meaning of the act of communication has to satisfy the methodological requirement of compositionality and utilizes a concept of pragmatic compositionality for meaning representations, called *meaning mergers* according to which the meaning of the act of communication is a function of the meaning of the words, the sentence structure, social and cultural assumptions triggered automatically, procedures of interpretation which rely on the properties of cognitive processes called cognitive defaults, and conscious pragmatic inference. Compositionality is a feature of a representation which is a merger of information coming from these diverse sources.[25] Viewed in this way, there is no priority given to sentence structure: sentence structure constitutes one of several sources of information about meaning and information it provides can be overridden, just as information derived from any other source can be overridden. There is no place in this model for a syntactic constraint.

The main point I am making here is this. We have a choice of definitions of the main object of analysis (main meaning) and one of the principles on which we should found our decision is the acceptance or rejection of the syntactic constraint. I would like to suggest that the decision is made by appealing to psychological considerations: the boundary between the primary meaning, which is the most salient meaning construed as that intended by the Model Speaker and recovered by the Model Addressee, and secondary meaning (implicatures that follow it) has to be psychologically real and empirically testable. Whether it uses the sentence as its skeleton or not need not concern us when we are in pursuit of the main communicated message. Note that neither Recanati's intuitively available and automatically processed *what is said* nor relevance-theoretic explicature are incompatible with this proposal in principle: they can be easily stripped of the requirement of the syntactic constraint. In fact, Carston's (2002) account of *ad hoc* concept construction is only one step from affecting some parts of syntactic structure of the sentence, just as it undermines the need for coded meaning. It is precisely by shunning psychological considerations that we become preoccupied with the syntax-pragmatics interface, to the detriment of what pragmatic theory should be really about. Frege repeatedly expressed his mistrust in syntactic categories, claiming that syntax of natural language can be misleading with regard to logical form.[26] It seems only natural to go all the way and allow for the cases where syntax has such a small role to play in the representation of the truth-conditional content that it does not even provide a skeleton. Moreover, there is substantial experimental evidence in support of the claim that the main, most salient meaning is frequently an implicature: according to Sysoeva's experiments, for example, between 60 and 80 per cent of informants (depending on the language and culture) select implicatures as the main communicated meaning.[27]

The final question to address is how does this psychologically plausible unit, free from the syntactic constraint, fare with Frege's notion of a *judgment* (Frege 1879) or *thought* (Frege 1918–19)? Let us look again at the passage quoted in Section 2. In *Logic*, Frege says: "Logic is concerned with the laws of truth, not with the laws of holding something to be true, not with the question of how people think, but with the question of how they must think if they are not to miss the truth" (Frege 1897[1969]: 250).

The consequences for the semantics of natural language are as follows. When one takes a sentence with, say, indexical terms in it, such as "I am hungry," *being true* cannot apply to this sentence directly. If it did, we would have to say that the sentence is true for some speakers and false for others. Instead, the sentence expresses different thoughts ('Gedanken') when uttered by different persons. It is this thought that is true or false, not the sentence. The referent, the time, and place are provided by the thought, leaving "being true" as "placeless and timeless" (Frege 1893: 203). This view is fully respected in post-Gricean pragmatics. Fregean thoughts are not particular person's thoughts; a thought "needs no owner" (Frege 1918–19: 337) and one can agree that "[s]ince thoughts are not mental in nature, it follows that every psychological treatment of logic can only do harm" (Frege [1897]1969: 250). In the conceptual notation proposed in *Begriffsschrift*, the judgeable content is prefixed with a symbol for assertion (judgment) to form a unit which can be assessed for truth or falsehood. And the analysis of this unit corresponds to the analysis of thought. Thoughts are legitimate recipients of truth conditions. Not only do they rescue truth from temporal and spatial relativism, but they also provide a unit with assigned reference to indexical and other deictic expressions, disambiguated lexically and syntactically[28] to fulfill Grice's (1978) conditions for what is said.

More importantly, there is nothing to stop us using this concept more liberally in the spirit of radical pragmatics (see, e.g., Atlas 1977, 1979, 1989, 2005, 2006a). Thoughts (let us call them Neo-Fregean Thoughts) become then theoretical pragmatic constructs that have their semantic counterparts in underspecified semantic representations. When, for example, a sentence does not specify the scope of the negation operator, the corresponding Neo-Fregean Thought does. Thoughts are then compatible with enriched or "modulated" (Recanati 2004) propositions assumed in relevance theory or in Recanati's (2002, 2003, 2004) truth-conditional pragmatics. Such an appropriation of a Fregean thought is also what is modeled in merger representations of Default Semantics.

To conclude, I argued in this section that we need psychological arguments in choosing the unit which is to be modeled in the theory of meaning. Psychological considerations point towards a unit that is regarded as the main, most salient meaning intended by the speaker and recovered by the addressee – or Model Speaker and Model Addressee respectively, if we adopt the perspective assumed in Section 3.

7.   Psychologism and the contextualism–minimalism debate

From the discussion of the identified areas [1]–[4] we can build the following generalization. With respect to [1], Levinson's presumptive meanings and my default-semantic primary meanings both adopt the Model Speaker – Model Addressee perspective. For area [2], Levinson's theory adopts an overly local unit for pragmatic enrichments, while Default Semantics does not have an answer to the question as to at what stage exactly the inference or automatic enrichment take place and prefers to look at these processes globally, as if they were post-propositional.

Table 1. Presumptive meanings and primary meanings w.r.t. [1]–[4]

	Levinson's (2000) presumptive meanings	Jaszczolt's (2005) primary Meanings
[1]	√ +P	√ +P
[2]	× +P	× +P
[3]	? +P	? +P
[4]	× +P	√ +P

Regarding [3], they are both lacking in empirically implementable definitions of what counts as automatic enrichment vis-à-vis conscious pragmatic inference. In [4], I argued that the correct way to construe a psychologically real unit of meaning is to exorcise the syntactic constraint – the move that was made in Default Semantics but not in other contextualist accounts. The analyses of all these areas in both frameworks make at least a modest use of psychological considerations of utterance processing. I summarize these conclusions in Table 1, where √ stands for a solution that was here argued to be adequate, × stands for a solution that was found lacking in adequacy, ? stands for no solution given, and +/- P for the need for or making use of psychological considerations.

Psychologism seems vindicated: where several solutions are available for an area, such as (I)–(III) for area [1], I have shown that making the choice involves psychological arguments. Where no satisfactory solution has been reached, I have argued that this is so due to shunning arguments from processing. A question arises at this point: are there, or can there in principle be theories of meaning that can be classified as -P? The obvious candidates are minimalist semantic accounts of Borg (2004), Cappelen and Lepore (2005), or even more so Bach's (2006) "radical minimalism". In what follows I briefly address the question as to whether they contain a construct of sentence meaning that is truly -P, leaving psychology outside the realm of truth-conditional analysis.

In the past three years, arguments against psychologism in the theory of meaning have been greatly aided by the revival of the so-called minimalist approaches in the form of Borg's (2004) *minimal semantics* and Cappelen and Lepore's (2005a, 2005b) *insensitive semantics*. After three decades of radical pragmatics where underdetermined semantic representation has been thought of as further developed, enriched, modulated, and so forth, by the result of pragmatic processing,[29] the traditional view of a clear semantics-pragmatics distinction is experiencing a revival in a variety of directions. Borg (2004) claims that pragmatic considerations are separate from semantics where the latter is a separate module and concerns the logical form as a property of expressions themselves. Non-demonstrative inference is to be kept apart from semantics which confines itself to formal, deductive operations (see p. 8). The notion of truth conditions is equally minimal on her account. For example, a demonstrative "that" in "That is red" is not to be filled with a referent within the semantics: context and the verification of the sentence in a situation are outside of the inquiry. The truth condition is, as she says, liberal: "that" figures in it as a singular term referring to a contextually salient object, whatever this object happens to be in this particular context. In this way psychologism is exorcized by decree (but I return below to the question as to whether it is really absent there). In a similarly minimalist account, Cappelen and Lepore overtly refer to Frege's ban on psychologism in the theory of meaning (Cappelen and Lepore 2005a: 152–153) when they suggest that semantic theory is to identify that content of the sentence that is shared across contexts. The minimal, semantic content is contrasted with what is said (the communicated content) and the first only allows for the filling in of context-sensitive terms, confined to very few classes of expressions, such as demonstratives and indexicals.[30] The latter, what is said, is radically unconstrained: a sentence can express indefinitely many propositions. They call this view Speech Act Pluralism.

The problem with these two ways of banning psychologism by staying close to the sentence meaning is what Bach (2004, 2005, 2006) calls their Propositionalism. Once we adopt a proposition as the object of study of semantic theory and accept that every indexical-free sentence must express a proposition, we have already mixed up psychological and formal considerations. As I demonstrated elsewhere (Jaszczolt 2007), Borg (2004), while attempting to build an autonomous, formal and modular semantics, resorts to *utterance* meaning for some types of expressions (see her discussion of "It is raining") in order to fulfill the requirement of having a proposition, a unit with (however minimal) truth conditions. She also resorts to psychologism while arguing for the co-existence of contextualism of the relevance-theory type and her minimalism.[31] Cappelen and Lepore's recourse to pragmatics in filling in the context-dependent expressions without which propositionality cannot be achieved is also a signal of their, however small, making allowances for context, processing, and psychological factors.

We are compelled to conclude that just as Grice's and relevance-theoretic programs cannot co-exist because they both resort to claims about cognition and some degree of psychologism, so neither of the two minimalist stances discussed in this section can happily co-exist with contextualism for exactly the same reason. While Cappelen and Lepore's account can be made compatible with contextualism on methodological grounds because it contains a very clear list of such context-dependent expressions[32] which we could treat as exceptions for an otherwise truly minimalist account, Borg's arguments make her more contextualist than her professed orientation would justify.

8.   Concluding remarks

It remains to be seen whether the solution is to ban propositions in order to ban psychologism, or to retain propositions and admit some modest dose of psychologism. In this paper I argued for the latter because a proposition-free semantics that is properly formally constrained and compositional is for me inconceivable. Exorcising propositions means exorcising truth conditions. We need a more detailed proposal from anti-propositionalists to challenge the foundations of Tarski, Montague, and of currently very successful dynamic semantic approaches. My contribution therefore does not end the investigation but rather contributes a modest interim conclusion that there is no third alternative: no proposition-based, truth-conditional theory of meaning without at least moderate psychologism. In particular, I argued that psychologism in truth-conditions-based pragmatic theory is necessary in order to formulate food for experimentation on at least the following fronts: [1] the perspective which should be adopted: that of the speaker, the addressee, or a Model Speaker – Model Addressee interaction; [2] the unit on which the pragmatic inference or default enrichment operate; [3] the definition and delimitation of default interpretations vis-à-vis conscious pragmatic inference; and [4] what counts as the main meaning to be modeled. This considerably narrows down the playing field.

Notes

1.   I am grateful to Aly Pitts, David Cram, Mikhail Kissine, and the participants of Istvan Kecskes and Jacob Mey's panel on intentions and common ground at the 10th International Pragmatics Conference for the discussion of various aspects of this paper. I am also indebted to Jay Atlas for drawing the contentious issue of psychologism in pragmatic theory to my attention through his recent papers.
2.   But see Travis (2006: 125-6) who tentatively suggests that taking *any* stance, including Fregean, on how logical laws apply to thinking subjects may constitute a form of psychologism.
3.   See Frege's letters to Husserl (Frege [1906]1976) and his review of Husserl's *Philosophy of Arithmetic I* (Frege 1894).
4.   *Pace* Travis 2006.
5.   On Frege's function/argument analysis of judgments see also, e.g., Baker and Hacker 2003, Green 2006, and Stalmaszczyk 2006.
6.   The literature on the semantics-pragmatics boundary issue is ample. For an overview see e.g. Horn 2006 and Jaszczolt 2002, forthcoming.

7. I attend to the notion of common ground and cultural defaults in Section 4 while discussing my area [2].

8. See Green 2003, Section 2 where she points out the connection between the semantics of sentences and the structure of perceptual judgment, making the case for the priority of ontological categories over those of syntax. Nota bene, she says that "[l]anguage ... records and conveys the results of perceptual judgements" (213), showing thereby that the kind of psychologism Frege bans is not the psychological reality of his analysis of judgements but rather the need for talking about individual minds. I come back to this issue at the end of this section.

9. See Section 1 above.

10. For some results of empirical studies of salient meanings see Giora (2003) and her Graded Salience Hypothesis. See also Kecskes (2008) on individual and collective salience.

11. Common ground is not a primitive concept on this construal and in a theory of communication has to be accounted for in terms of the effects of conscious pragmatic inference and (various categories of) default meanings. I am grateful to David Cram and to the participants of Istvan Kecskes and Jacob Mey's IPrA conference panel on intentions and common ground (Göteborg, July 2007) for the discussion of this point.

12. Evidence and theoretical arguments in support of the existence of default interpretations are indeed compelling. See, e.g., Horn 2004, Levinson 2000, Asher and Lascarides 2003, Jaszczolt 2005a, Bach 1984, Veltman 1996, Giora 2003 for various aspects of, and approaches to, default meanings. See also Jaszczolt 2006a for an overview of the seminal accounts of defaults in semantics and pragmatics.

13. See Horn 2006, Saul 2002, Bach 2001, and Geurts 1998.

14. See Section 1.

15. Compounds can exhibit different orthographic conventions, they can be written jointly, separately, or hyphenated. The fact that the examples in (4) are written separately is immaterial.

16. The literature on this topic is vast and growing rapidly. See, e.g., articles in Noveck and Sperber 2004, Katsos 2007.

17. I discussed the defeasibility problem at length in Jaszczolt 2005a while demonstrating the advantages of global, post-propositional defaults of Default Semantics, so will resort to repeating it only briefly.

18. Default Semantics subscribes to what Bach (2006) calls Propositionalism, a view that the proper object of study of a theory of meaning is a proposition and that that proposition is recovered from the sentence and the context – here in the form of default meanings and pragmatic inference. On an alternative view, represented by Bach, a semantic minimalism that is even more minimal than minimal propositions in that it is not even a task of semantics to deliver

truth conditions, see Bach 2004. On different forms of minimalism in semantics see Section 6 below and Jaszczolt 2007.

19. This construal is in fact compatible with the general assumptions of relevance theory (e.g., Sperber and Wilson 1995) and Recanati's truth-conditional pragmatics (e.g., Recanati 2004). In Sysoeva and Jaszczolt (2007) we develop an argument demonstrating that contextualist approaches to meaning need not be bound by the syntactic constraint, i.e., that the primary meaning (explicature, what is said, or even Bach's impliciture) need not be construed as a development of the logical form of the uttered sentence.

20. I use the acronym "NP" in the theory-neutral sense.

21. We are disregarding for the purpose of this argument the history of lexicalization that, according to one influential theory, proceeds through the stages PCI>GCI>SM (where "PCI" stands for particularized conversational implicatures and "SM" for semantic meaning). See Traugott 2004.

22. See also Kecskes (2008) on collective salience.

23. The interpretation on which "Leonardo" is identified as short for "Leonardo da Vinci" would be plausible if the proper name referred, say, to Leonardo's paintings and thereby "working with Leonardo" could mean "working with Leonardo's paintings." In the context of the interview in (17) it could also stand for an actor playing the part of Leonardo da Vinci.

24. The terminology and minor differences between contextualist Gricean views are irrelevant for this purpose.

25. The view that meaning construction draws on various sources is not new. It can be traced back to Husserl's idea of objectifying acts from early 20th century (see my account of Husserl's vehicles of thought in Jaszczolt 1999). See also Kecskes' (2008) attempt to reconcile invariant meanings with Wittgenstein's eliminativism, modeling the mind as a "pattern recogniser/builder" but also a "rule-following calculator."

26. See Green 2003, especially 216–217 for an excellent discussion of the problems with deriving ontology from syntax. She convincingly argues that Frege's ontological categories of *object* and *concept* are prior in the order of explanation to the syntactic categories of *singular term* and *predicate*.

27. See Sysoeva and Jaszczolt 2007. For similar results obtained by different methods see Nicolle and Clark 1999 and Pitts 2005.

28. See Jaszczolt 1999, Chapter 1 on ambiguity in semantics.

29. See, e.g., Cole 1981, Atlas 1979, Kempson 1975, 1986, and for an overview Recanati 2005 and Jaszczolt forthcoming.

30. I am confining this discussion to the forms of minimalism that adhere to Propositionalism (see footnote 18). I also exclude what I called elsewhere "pseudo-minimalism" (Jaszczolt 2007) where syntactic slots are liberally postulated for various forms of pragmatic enrichment. See Stanley (2002), Stanley and Szabó (2000) and King and Stanley (2005). Minimalism is

achieved there by making the syntactic structure unjustifiably rich and unpronounced. To compare, Recanati 2005 has a more liberal category of minimalisms under which this stance comfortably fits: on Recanati's (2005: 176) definition, to count as a minimalist one has to maintain that "no contextual influences are allowed to affect the truth-conditional content of an utterance unless the sentence itself demands it."

31. See Borg (2004: 243) and her appeal to children's and philosophers' intuitions about minimal propositions. Minimal propositions, as units of autonomous, formal, and modular semantics are not supposed to have any psychological reality, or at least their role in processing should be irrelevant. In contextualism, minimal propositions do not exist. They don't even make sense as theoretical constructs. For contextualists, every proposition requires pragmatic enrichment (modulation). See Recanati (2005: 179–180).

32. Their tentative list contains personal pronouns, demonstrative pronouns, adverbials such as "here", "there", "now", "two days ago", adjectives "actual" and "present", and temporal expressions. See Cappelen and Lepore (2005a: 144).

References

Asher, Nicholas and Alex Lascarides
    2003   *Logics of Conversation*. Cambridge: Cambridge University Press.
Atlas, Jay David
    1977   Negation, ambiguity, and presupposition. *Linguistics and Philosophy* 1. 32–36.
    1979   How linguistics matters to philosophy: Presupposition, truth, and Meaning. In *Syntax and Semantics*. Vol. 11: *Presupposition*, David A. Dinneen and Choon-Kyn Oh (eds.), 265–281. New York: Academic Press.
    1989   *Philosophy without Ambiguity: A Logico-Linguistic Essay*. Oxford: Clarendon Press.
    2005   *Logic, Meaning, and Conversation: Semantical Underdeterminacy, Implicature, and Their Interface*. Oxford: Oxford University Press.
    2006a   A personal history of linguistic pragmatics 1969–2000. Paper presented at the *Jay Atlas-Distinguished Scholar Workshop*, University of Cambridge.
    2006b   Remarks on F. Recanati's *Literal Meaning*. Unpublished Manuscript.

in press   Meaning, propositions, context, and semantical underdeterminacy. In
            Essays on Insensitive Semantics, Gerhard Preyer (ed.), Oxford:
            Oxford University Press.
Bach, Kent
1984   Default reasoning: Jumping to conclusions and knowing when to
            think twice. *Pacific Philosophical Quarterly* 65. 37–58.
1994   Semantic slack: What is said and more. In *Foundations of Speech
            Act Theory: Philosophical and Linguistic Perspectives*, Savas L.
            Tsohatzidis (ed.), 267–291. London: Routledge.
1995   Remark and reply. Standardization vs. conventionalization.
            Linguistics and Philosophy 18. 677–686.
2001   You don't say? *Synthese* 128.15–44.
2004   Minding the gap. In *The Semantics/Pragmatics Distinction*, Claudia
            Bianchi (ed.), 27–43. Stanford: CSLI Publications.
2005   Context *ex Machina*. In *Semantics versus Pragmatics*, Zoltan
            Gendler Szabó (ed.), 15–44. Oxford: Clarendon Press.
2006   The excluded middle: Semantic minimalism without minimal
            propositions. Unpublished paper.
Borg, Emma
2004   *Minimal Semantics*. Oxford: Oxford University Press.
Cappelen, Herman and Ernest Lepore
2005a   *Insensitive Semantics: a Defense of Semantic Minimalism and
            Speech Act Pluralism*. Oxford: Blackwell.
2005b   A tall tale: In defense of Semantic Minimalism and Speech Act
            Pluralism. In *Contextualism in Philosophy: Knowledge, Meaning,
            and Truth*, Gerhard Preyer and Georg Peter (eds.), 197–219. Oxford:
            Clarendon Press.
Carston, Robyn
1988   Implicature, explicature, and truth-theoretic semantics. In *Mental
            Representations: The Interface Between Language and Reality*, Ruth
            M. Kempson (ed.), 155–181. Cambridge: Cambridge University
            Press.
2002   *Thoughts and Utterances: The Pragmatics of Explicit
            Communication*. Oxford: Blackwell.
Cole, Peter (ed.)
1981   *Radical Pragmatics*. New York: Academic Press.
Frege, Gottlob
1879a   'Begriffsschrift, eine der arithmetischen nachgebildete
            Formelsprache des reinen Denkens'. Halle: L. Nebert. [Conceptual
            notation: A formula language of pure thought modelled upon the
            formula language of arithmetic]. In *Conceptual Notation and
            Related Articles*, Terrell Ward Bynum (ed.), 101–203. 1972. Oxford:
            Oxford University Press.

1879b 'Begriffsschrift, eine der arithmetischen nachgebildete Formelsprache des reinen Denkens'. Halle: L. Nebert. [*Begriffsschrift*: a formula language of pure thought modelled on that of arithmetic]. In *The Frege Reader*, Michael Beaney (ed.), 47–78. 1997. Oxford: Blackwell.

1884a *Die Grundlagen der Arithmetik, eine logisch mathematische Untersuchung über den Begriff der Zahl*. Breslau: W. Koebner. Trans. John L. Austin. [*The Foundations of Arithmetic: A Logico-Mathematical Enquiry into the Concept of Number*]. 1953. Oxford: B. Blackwell.

1884b *Die Grundlagen der Arithmetik, eine logisch mathematische Untersuchung über den Begriff der Zahl*. In *The Frege Reader*, Michael Beaney (ed.), 84–91. 1997. Oxford: Blackwell.

1893 *Grundgesetze der Arithmetik*. Vol.ume 1. In *The Frege Reader*, Michael Beaney (ed.), 194–208. 1997. Oxford: Blackwell.

1894 Review of E. G. Husserl, *Philosophie der Arithmetik I* [Philosophy of Arithmetic I]. *Zeitschrift für Philosophie und philosophische Kritik* 103. In G. Frege. *Collected Papers on Mathematics, Logic, and Philosophy*, Brian McGuinness (ed.), 195–209. 1984. Oxford: Blackwell.

1969 Reprint. *Logic*. In: *Nachgelassene Schriften*. Hamburg: Felix Meiner. In: 1979. *Posthumous Writings*. Oxford: Blackwell. Sections 1 (Introduction) and 2 (Separating a thought from its trappings) In *The Frege Reader,* Michael Beaney (ed.), 227–250. 1997. Oxford: Blackwell. 1897.

1976 Reprint. *Letters to Husserl, 1906*. In: *Wissenschaftlicher Briefwechsel*. 1976. Hamburg: Felix. 1906.

1980 Meiner. In: G. Frege. *Philosophical and Mathematical Correspondence*. Brian McGuinness (ed.), 66–71. Oxford: Blackwell. In *The Frege Reader,* Michael Beaney (ed.), 301–307. 1997. Oxford: Blackwell.

1997a. Reprint. 'Der Gedanke'. *Beiträge zur Philosophie des deutschen Idealismus* I. [Thoughts (Part I of *Logical Investigations*)] In G. Frege. *Collected Papers on Mathematics, Logic, and Philosophy*, Brian McGuinness (ed.), 1984. Oxford: Blackwell. In *The Frege Reader*, Michael Beaney (ed.), 325–345. 1997. Oxford: Blackwell. 1918–1919.

Grice, H. Paul

1989a Reprint. Logic and conversation. In *Syntax and Semantics* 3, Peter Cole and Jerry L. Morgan (eds.), New York: Academic Press. Reprinted in: H. Paul Grice. *Studies in the Way of Words*. Cambridge, MA: Harvard University Press. 1975. 22–40.

1989b   Reprint. Further notes on logic and conversation. In *Syntax and Semantics* 9, Peter Cole (ed.), New York: Academic Press. Reprinted in *Studies in the Way of Words*, H. Paul Grice. 41–57. Cambridge, MA: Harvard University Press. 1975.

Geurts, Bart
1998   Scalars. In *Lexikalische Semantik aus kognitiver Sicht*, Petra Ludewig and Bart Geurts (eds.), 95–117. Tübingen: Gunter Narr.

Giora, Rachel
2003   *On Our Mind: Salience, Context, and Figurative Language*. Oxford: Oxford University Press.

Horn, Laurence Robert
2004   Implicature. In *The Handbook of Pragmatics*, Laurence Horn and Gregory Ward (eds.), 3–28. Oxford: Blackwell.
2006   The border wars: A neo-Gricean perspective. In *Where Semantics Meets Pragmatics: The Michigan Papers*, Klaus von Heusinger and Kenneth Turner (eds), 21–48. Oxford: Elsevier.

Jaszczolt, Katarzyna M.
1999   *Discourse, Beliefs, and Intentions: Semantic Defaults and Propositional Attitude Ascription*. Oxford: Elsevier Science.
2002   *Semantics and Pragmatics*. London: Longman.
2005a   *Default Semantics: Foundations of a Compositional Theory of Acts of Communication*. Oxford: Oxford University Press.
2005b   Review of E. Borg, *Minimal Semantics*. *Journal of Linguistics* 41. 637–642.
2006a   Defaults in semantics and pragmatics. In *Stanford Encyclopedia of Philosophy*. Edward N. Zalta (ed.), accessed at http://plato.stanford.edu/contents.html
2006b   Meaning merger: Pragmatic inference, defaults, and compositionality. *Intercultural Pragmatics* 3 (2): 195–212.
2007   On being post-Gricean. In *Interpreting Utterances: Pragmatics and Its Interfaces. Essays in Honour of Thorstein Fretheim*, Randi A. Nilsen, Nana Amfo and Kaja Borthen (eds.), 21–38. Oslo: Novus.
in press   Semantics and pragmatics: The boundary issue. In *Semantics: An International Handbook of Natural Language Meaning*, Klaus von Heusinger, Paul Portner and Claudia Maienborn. Berlin/New York: Mouton de Gruyter.

Kamp, Hans, Josef van Genabith and Uwe Reyle
in press   Discourse Representation Theory. In *Handbook of Philosophical Logic*. Second edition.: Dov M. Gabbay and Franz Guenthner (eds.).
1993   *From Discourse to Logic: Introduction to Modeltheoretic Semantics of Natural Language, Formal Logic and Discourse Representation Theory*. Dordrecht: Kluwer.

Katsos, Napoleon
  2007   The semantics/pragmatics interface from an experimental
             perspective: The case of scalar implicature. Unpublished manuscript.
Kecskes, Istvan
  2008   Dueling contexts: A dynamic model of meaning. *Journal of
             Pragmatics*. 40 (3): 385–406.
Kempson, Ruth M.
  1975   *Presupposition and the Delimitation of Semantics*. Cambridge:
             Cambridge University Press.
  1986   Ambiguity and the semantics-pragmatics distinction. In *Meaning
             and Interpretation*, Catherine Travis (ed.), 77–103. Oxford: B.
             Blackwell.
King, Jeffrey C. and Jason Stanley
  2005   Semantics, pragmatics, and the role of semantic content. In
             Semantics versus Pragmatics, Zoltan Gendler Szabó (ed.), 111–164.
             Oxford: Clarendon Press.
Levinson, Stephen C.
  1995   Three levels of meaning. In *Grammar and Meaning. Essays in
             Honour of Sir John Lyons*, F. R. Palmer (ed.), 90–115. Cambridge:
             Cambridge University Press.
  2000   *Presumptive Meanings: The Theory of Generalized Conversational
             Implicature*. Cambridge, MA: MIT Press.
Nicolle, Steve and Billy Clark
  1999   Experimental pragmatics and what is said: A response to Gibbs and
             Moise. *Cognition* 69. 337–354.
Noveck, Ira A. and Dan Sperber (eds.)
  2004   *Experimental Pragmatics*. Basingstoke: Palgrave Macmillan.
Pitts, A.
  2005   Assessing the evidence for intuitions about *what is said*. University
             of Cambridge. Unpublished manuscript.
Recanati, François
  1991   Reprint. The pragmatics of what is said. *Mind and Language* 4. In
             Pragmatics: A Reader, Steven Davis (ed.), 97–120. Oxford: Oxford
             University Press. 1989.
  2002   Unarticulated constituents. *Linguistics and Philosophy* 25: 299–345.
  2003   Embedded implicatures. http://jeannicod.ccsd.cnrs.fr/documents.
  2004   *Literal Meaning*. Cambridge: Cambridge University Press.
  2005   Literalism and contextualism: Some varieties. In *Contextualism in
             Philosophy: Knowledge, Meaning, and Truth*, Gerhard Preyer and
             Georg Peter (eds.), 171–196. Oxford: Clarendon Press.
Saul, Jennifer M.
  2002   What is said and psychological reality; Grice's project and relevance
             theorists' criticisms. *Linguistics and Philosophy* 25. 347–372.

Sperber, Dan and Deirdre Wilson
    1995   Reprint. *Relevance: Communication and Cognition.* Oxford:
            Blackwell. 1986.
Stalmaszczyk, Piotr
    2006   Fregean predication: between logic and linguistics. *Research in
            Language* 4. 77–90.
Stanley, Jason
    2002   Making it articulated. *Mind and Language* 17. 149–168.
Stanley, Jason and Zoltan Gendler Szabó
    2000   On quantifier domain restriction. *Mind and Language* 15. 219–261.
Sysoeva, Anna and Katarzyna Jaszczolt
    2007   Composing utterance meaning: An interface between pragmatics and
            psychology. Paper presented at the 10th International Pragmatics
            Conference, Göteborg, July 2007.
            http://www.cus.cam.ac.uk/~kmj/pwpt.html
Traugott, Elizabeth C.
    2004   Historical pragmatics. In *The Handbook of Pragmatics*, Laurence
            Horn and G. Ward (eds.), 538–561. Oxford: Blackwell.
Travis, Charles
    2006   Psychologism. In *The Oxford Handbook of Philosophy of Language*,
            Ernest Lepore and Barry C. Smith (eds.), 103–126. Oxford:
            Clarendon Press.
Van Eijck, Jan and Hans Kamp
    1997   Representing discourse in context. In *Handbook of Logic and
            Language*, Johan Van Benthem and Alice Ter Meulen (eds.), 179–
            237. Amsterdam: Elsevier Science.
Veltman, Frank
    1996   Defaults in update semantics. *Journal of Philosophical Logic* 25.
            221–261.

The place of intention in the interactional achievement of implicature

Michael Haugh

1. Introduction

It is commonly assumed in linguistic pragmatics that the communication of implicatures involves the addressee making inferences about the intention(s) of the speaker. The notion of intention generally invoked in pragmatics can be broadly defined as "an element 'inside' a person which motivates him/her to act in certain ways" (Nuyts 2000: 1), which builds upon the intuitive understandings of *intention* that ordinary speakers of English for the most part share (Gibbs 1999: 23; Malle 2004; Malle and Knobe 1997). In other words, the notion of intention commonly presumed in pragmatic processing arguably encompasses an *a priori* mental construct where "some individual has in mind some situation, not yet actualized, along with a disposition to prefer that the situation be actualized" (Mann 2003: 165).

This intention-based view of communication and implicature finds its roots in Grice's (1957) seminal work on meaning and speaker intention, in particular, Grice's insight that a speaker meant$_{nn}$ something by x if and only if S "intended the utterance of x to produce some effect in an audience by means of the recognition of this intention" (Grice 1957: 385). According to this view, successful communication can only be achieved when the hearer recognizes the speaker's intention to mean something by the saying of x. It is the inherent reflexivity of the Gricean notion of speaker intention that has perhaps prompted the move, though not necessarily intended by Grice himself as carefully argued by Arundale (1991), from speaker (intended) meaning to communication in general within various approaches in pragmatics. A number of different approaches to communication in pragmatics have thus developed, building in various ways on Grice's work, including Gricean and neo-Gricean Pragmatics (Bach 1987, 2006: 23; Dascal 2003: 22–23; Grice [1967]1989; Horn 2004: 3; Levinson 1983: 16–18; 2000: 12–13, 2006a: 87, 2006b: 49; Recanati 1986), Relevance Theory (Breheny 2006: 97; Carston 2002: 377; Sperber and Wilson 1995: 194–

195), Speech Act Theory (Searle 1969, 1975), and Expression Theory (Davis 2003: 90). While these various approaches differ to some extent, both in how they conceptualize speaker intentions, and in regards to what constraints they posit as influencing the inferential processes leading to hearers attributing those intentions to speakers, the received view of implicature, and more broadly communication, is that they crucially involve inferences about speaker intentions. And although it is rarely commented upon in the literature, the intention-based view of implicature, and more broadly communication which predominates in pragmatics, invariably conceptualizes these intentions as *a priori*, conscious mental states of individual speakers (Gibbs 1999: 23, 2001: 106; Mann 2003: 165).

One challenge facing the intention-based view of implicature, however, is the view that meanings *emerge* through joint, collaborative interactions between speakers and hearers rather than being dependent on inferences about speaker intentions (Arundale 2008; Drew 2005: 171; Hopper 2005: 149; Schegloff 1991: 168; 1996: 183–184). If one considers the manner in which implicatures arise in discourse, for instance, the relationship between (speaker) intentions and what is implied becomes somewhat more complex (Haugh 2007), if not outright problematic (Gauker 2001, 2003; Marimaridou 2000).

A second key challenge facing the received view is the claim that many implicatures are not dependent on the recognition of speaker intentions, but rather are a matter of *conventionality* (Davis 1998: 190; 2007: 1671). This position is also implicit (although not acknowledged) in Levinson's (1995, 2000) notion of default inferences, proposed within his broader Theory of Generalized Conversational Implicature, namely, that speaker intentions underdetermine generalized or 'default' implicatures. Burton-Roberts (2006: 8) has thus recently made the claim that generalized conversational implicatures arise independently of any intention of the speaker. On the other hand, the existence of default inferences continues to be disputed by Relevance Theorists and others (Bezuidenhout 2002; Carston 1995, 1998, 2002). The relationship between implicatures, (speaker) intentions, and conventionality is clearly more complex than has been commonly assumed to date.

In this chapter, it is first argued that this speaker intention-based view of implicature does not sufficiently acknowledge the inherent temporal, ontological and epistemological ambiguity of intentions in discourse. It is then suggested that an approach which conceptualizes meaning in communication as conjointly co-constituted, that is, as a non-summative and emergent interactional achievement (Arundale 1999, 2005; Arundale and Good 2002), allows for an analysis of both nonce and default implicatures in interaction without recourse to the notion of intention. This analysis is predicated on an emergent notion of intentionality in the sense of the presumed "aboutness" or "directedness" of utterances in discourse (Duranti 2006: 36; Nuyts 2000: 2–3; Searle 1983: 1), to which speakers are held contingently accountable by interactants (Garfinkel 1967; Heritage 1984). In the final section, it is proposed that the notion of intention need only be invoked in particular instances where it emerges as a *post facto* construct salient to the interactional achievement of implicatures.

2. The ambiguous status of intention(s) in discourse

While the view that hearers make inferences about *a priori* speaker intentions in order to generate implicatures dominates theories in pragmatics, most if not all the examples adduced to support such a view have inevitably been at the utterance level. If one considers the manner in which implicatures arise in discourse, however, the picture becomes somewhat more complex. In the following sections, the ambiguity that surrounds the temporal, ontological and epistemological status of the intentions underlying implicatures in discourse is examined in more detail.

2.1. Temporal ambiguity

Although it is difficult to find a clear statement of the position of theorists in relation to the temporal status of intentions underlying implicature, it appears that (the hearer's inferences about) the *a priori* intentions of the speaker are commonly assumed to underlie the generation of implicature in pragmatics. An *a priori* intention is closest to what we normally understand *intention* to be in the intuitive folk sense, as a plan or aim formulated before an action (including linguistic and paralinguistic behavior) by the speaker. More recently, it has also been proposed under the umbrella of Relevance Theory, drawing upon Bratman's (1987, 1990) work on "future-directed intentions" that (*a priori*) higher-order intentions may constrain the production and interpretation of discourse, as well as single utterances within that discourse (Taillard 2002: 189–190; Ruhi 2007).[1]

However, intention can be used in multiple ways to refer to different points in time in relation to specific utterances in discourse. For instance, it has also been claimed that the intentions underlying actions do "not terminate with the onset of action but continues until the action is completed" (Pacherie 2000: 403), or what Searle (1983: 93) terms an "intention-in-action." Indirect support for such a view can be found in an experimental study in which the researcher asked subjects to listen to an extended interaction involving a request by one person and an eventual refusal of this request by the other (Meguro 1996). It was found that the subjects' assessments of whether the request was being refused initially started with not knowing, but quickly shifted to thinking the request was being accepted, before finally ending with a gradual shift towards thinking it was being refused. While potential differences between the understandings of observers and participants in conversation (Clark 1996; Holtgraves 2005) mean such results should be treated with due caution, it is apparent from this study that inferences about the intentions of others can and do change over time.

Moreover, it has been pointed out that intentions themselves can be topicalized in interaction by communication theorists (Arundale and Good 2002: 127–128; Buttny 1993; Buttny and Morris 2001; Jayyusi 1993), conversation analysts (Heritage 1984, 1988; 1990/91; Moerman 1988), social psychologists (Gibbs 1999, 2001; Malle 2001, 2004), and discursive psychologists (Edwards 1997: 107–108, 2006: 44, 2008; Edwards and Potter 2005: 243–244; Locke and Edwards 2003; Potter 2006: 132). Intentions may be invoked *post facto* by interactants in accounting for their actions or to question the actions of others. Edwards and Potter (2005), for instance, treat intentions as participant resources that "attend to matters local to the interactional context in which they occur and…attend reflexively to the speaker's stake or investment in producing those descriptions" (246). Intention can thus be conceptualized as a *post facto* construct that is jointly attributed by interactants to utterances and longer sections of discourse in accounting for violations of normative expectations.

While analyzing speaker intentions from this *post facto* perspective does not in itself negate an analysis of *a priori* speaker intentions, it does indicate that in theorizing implicatures the analyst is often faced with an inherent ambiguity in the temporal status of the intentions involved. Thus while introspection allows us to assume that *a priori* intentions do presumably exist at times, careful analysis of the emergence of implicatures in actual interactional data suggests that the relationship between such *a priori* intentions and implicatures is difficult to establish. In the following example, as previously argued by Heritage (1990/1991), it is somewhat difficult to determine at which point in time in an interaction an intention was formed and indeed the scope of this intention.

(1) (S has just called G up on the telephone)

```
1    G:     ...d'ju see me pull up?=
2    S:     =.hhh No:. I w'z trying you all day.=en the
3           line w'z busy fer like hours.
4    G:     Ohh:::::, ohh:::::, .hhhhh We::ll, hhh I'm
5           g'nna c'm over in a little while help yer
6           brother ou:t
7    S:     Goo[:d
8    G:        [.hh Cuz I know he needs some he::lp,
9    S:     .hh Ye:ah. Yes he'd mention'that tihday.=
10   G:     =Mm hm.=
11   S:     =.hh Uh:m, .tlk .hhh Who wih yih ta:lking to.
```
(Heritage 1990/1991: 317)[2]

There are at least two possible interpretations of this interaction according to Heritage (1990/1991). The first interpretation is that the utterance in lines 2–3 was an attempt at fishing as to whom G was talking to on the phone which failed, and thus was followed up in line 11 by a second more explicit attempt to find out (Pomerantz 1980). In other words, in lines 2–3 we can see an implicature (something like "Who were you talking to on the phone?") could have arisen from S's utterance, but G's subsequent response does not indicate that such an inference has been made, so this implicature is not interactionally achieved here. The "potential implicature" arising from S's utterance lines 2–3, however, is made explicit by the utterance in line 11, which may lead to a revision by G of her interpretation of S's utterance in lines 2–3.

An alternative interpretation, however, is that S's utterance in lines 2–3 was an explanation as to why S was calling G, since G's prior utterance in line 1 could be interpreted as seeking an account for why S came to be calling her. G's alleged non-response to the implicature now appears as an appropriate response to S's explanation in lines 2–3. In referring to S's repeated attempts to reach G, S could instead be interpreted as invoking G's earlier undertaking to help S's younger brother with some task. The choice between these two interpretations creates issues for an intention-based account of implicature as noted by Heritage (1990/ 1991).

> If S's initial turn was not intended as a "fish" but rather as an account, it is still possible that its failure to elicit the information it inadvertently solicits had the effect of raising a previously unspoken and unintended issue to consciousness - thus triggering the subsequent explicit question. Complex questions arise here about specifying an exact moment at which we may claim that S formed an "intention" to find out who G was talking to (318).

In this situation, then, even if S had the *a priori* intention to imply something like "Who were you talking to on the phone?" this intention in itself does not determine the course of the conversation, or indeed whether such an implicature arises or not. Moreover, to hold to the view that implicatures are determined by *a priori* speaker intentions gives rise to the counter-intuitive conclusion that this interaction involved miscommunication of the speaker's intentions. However, since the implicature attributed to S in making this utterance in line 2–3 could only have emerged through the unfolding of interdependent interpretings in the course of this interaction, it is questionable whether any analytical traction is to be gained from attempts to locate particular *a prior* intentions.

More recent analyses have also found similar difficulties in attempting to locate *a priori* intentions in interactional data (Drew 2005: 163; Hopper 2005: 149–150; cf. Sanders 2005: 63).The view that implicatures arise solely from *a priori* speaker intentions, and inferences about those intentions, does not therefore seem tenable in light of such analyses.

2.2. Ontological ambiguity

Another issue facing an intention-based approach to implicatures is that we need to carefully distinguish between the speaker's intention and the intention attributed to the speaker by the hearer (Mann 2003; Stamp and Knapp 1990), as what a speaker intends to imply is not necessarily what the hearer actually comes to understand (cf. Davis, 1998: 122). Various studies have indicated that how an implicature is understood by hearers is just as important as what the speaker might have intended in terms of what implicature arises in an interaction (Bilmes 1993; Cooren 2005; Sbisà 1992). For example, while I might intend to imply by asking the time when visiting a friend that I want to go home, this might also be legitimately understood by my friend as implying that I would like to stay for dinner, or simply that I want to keep track of the time for other reasons. In other words, an intention-based account of implicature raises the ontological issue of whose intentions we are really invoking. Most approaches to implicature appear to vacillate between a focus on speaker intentions and a focus on the hearer understandings of speaker intentions. However, this potential circularity is not limited to the hearers' understandings vis-à-vis the speakers' understandings of the speaker's intentions, but extends to question of the status of the analysts' understandings vis-à-vis the participants' understandings (Bilmes 1985), which has been for the most part neglected in analyses of implicatures. Another potential level of ambiguity in the analysis of implicatures arises from the distinction between individual speaker intentions and so-called we-intentions.

In regards to the first potential ontological ambiguity, on the one hand, Speech Act Theory (Searle 1969, 1975) and Expression Theory (Davis 1998, 2003) both privilege the speaker's understanding of his/her intentions as the basis for determining implicatures, while on the other, neo-Gricean Pragmatics (Grice [1967] 1989; Levinson 1983, 2000) and Relevance Theory (Sperber and Wilson 1995) appear to privilege the hearer's attribution of intentions to speakers in inferring implicatures. The first view that what speakers intend to mean can be equated with what is communicated is highly problematic in that it is not what the speaker intends per se that determines what is implied, but rather what the speaker says can be taken to imply according to the (sometimes generic) constraints of that context. For example, I may write a letter for a student for what I think is a reference for a philosophy job and write that he or she is good at typing with the intention of implying he is unsuitable for the job. However, if the reference turns out to be for a typing job, I fail to imply he or she is unsuitable for the job, and in fact do quite the opposite (Saul 2002: 230). In other words, the speaker's so-called "real" intentions are by necessity mediated by shared understandings about what can be legitimately implied by saying certain things.

However, the second view that implicatures arise from hearers making inferences about the speaker intentions is also problematic, as hearers can only know what speakers intend to imply if they know what others would see those speakers as implying (Bilmes 1986: 110). Yet as Bilmes (1986) points out, what others would see the speaker as implying is no more transparent to the hearer than what the speaker him/herself might have intended to imply. This highlights the ultimate circularity of the hearers' understandings of what has been implied vis-à-vis the speaker intentions.

"The crux of the matter is the speaker's intentions are displayed (insofar as they are displayed) because the utterance has meaning, and therefore, the meaning of the utterance cannot by wholly based on the speaker's displayed intention" (Bilmes 1986: 110). In other words, for a hearer to infer the speaker intention underlying a certain implicature, he or she must make recourse to expectations about what can or cannot be implied in that context, and so what is implied cannot *ipso facto* be based only on speaker intentions or hearer understandings of speaker intentions.

Recent work in psycholinguistics has also suggested hearers' inferences about speaker intentions may not play such a pivotal role as traditionally assumed. It has been argued, for example, that speakers consistently over estimate their ability to project specific intended meanings to addressees (Keysar 1994, 2000; Keysar and Henley 2002), and commonly do not use the ability to represent others' beliefs as much as they could (Keysar, Lin, and Barr 2003). It has been claimed from such experiments that speakers and hearers "routinely process language egocentrically, adjusting to the other's perspective only when they make an error" (Keysar, Barr, and Horton 1998: 46), or what is termed the tendency for egocentrism in language use' (Barr and Keysar 2005, 2006; Keysar 2007). This tendency also potentially undermines the notion that implicatures arise from the hearer's inferences about speaker intentions.

A second kind of ontological circularity that has received only passing attention in theories of implicature is the status of the analysts' understandings of speaker intentions vis-à-vis the participants. This circularity arises in the case of speaker intentions "when an analyst claims that talk shows evidence for the existence of a particular psychological state or process" and then goes on to "explain the production of that talk in terms of the existence of [that particular psychological state or process]" (Antaki, Billig, and Potter 2003: 13–14). According to Potter (2006), this kind of circularity can be seen in Drew's (2005) recent analysis of the putative speaker intentions involved in an interaction involving an invitation and anticipation of a refusal of that invitation by the first speaker. In this example, Emma invites Nancy to lunch after having inquired about what Nancy is presently doing.

(2) (Emma has called her friend Nancy on the phone)

```
1    E:      Wanna c'm do:wn 'av [a bah:ta] lu:nch w]ith me?=
2    N:                          [°It's  js]( )° ]
3    E:      =Ah gut s'm beer'n stu:ff,
```

```
4            (0.3)
5    N:      ↑Wul yer ril sweet hon: uh:m
6            (.)

7    E:      [Or d'y] ou'av] sup'n [else °(    )°
8    N:      [L e t- ]  I :    ]  hu.  [n:No: I haf to: uh call Roul's
9            mother, h I told'er I:'d call'er this morning. . .
(Drew 2005: 170)
```

Nancy's response to this invitation is slightly delayed (line 4), and consists of a *well*-prefaced appreciation in line 5. Emma then appears to anticipate that Nancy may be working towards declining the invitation, and so offers an account for this declination in line 7, which is subsequently confirmed by Nancy in lines 8–9. In other words, at this point in the interaction it appears Emma and Nancy have reached a common understanding that Nancy will probably not come for lunch. This is not, however, explicitly said and so can be understood as being implied. Drew (2005) goes on to suggest that in anticipating this (implied) refusal, Emma "*reads Nancy's mind*", attributing that intention to her" (170).[3] However, Potter (2006) argues that this attribution of intention by the analyst is not actually justified in Drew's analysis, and so such an assumption is inherently circular.

> Given the conventional nature of this relationship – given, that is, that the early elements are parts of the design of declinations – it is potentially circular to treat these elements as signs of an *intention* to do the act of declining. In effect, part of a declination is used as evidence of an intention to the do the declination which is, in turn (and here is the circle) evidence of the intention (Potter 2006: 135).

While the analyst's understanding is that Emma has attributed an intention to refuse her invitation to Nancy, the question is whether Emma herself has this same understanding, and indeed whether Nancy would infer that Emma has this understanding. In other words, is the analyst's understanding of the attribution of speaker intention consistent with that of the participants? It appears in this case that further work is required to avoid a divergence of the analysts' and participants' perspectives in relation to the possible attributions of speaker intentions that may or may not underlie the implicature that arises in this interaction.

Another potential level of ontological ambiguity in regards to the place of speaker intentions in generating implicatures is the distinction between individual and we-intentions, which proposes that certain actions involve an intention that is shared by participants (Searle 1990, 2002; Tuomela and Kaarlo 1988; Tuomela 2005).[4] Tomasello, Carpenter, Call, Behne and Molle (2005), for instance, define we-intention(ality) as "collaborative interactions in which participants have a shared goal (shared commitment) and coordinated action roles for pursuit of shared goal" (680). Searle (1990, 2002), among others, also argues that we-intentions cannot be regarded as the summation of individual intentional behavior, and so cannot be reduced to sets of individual intentions. The notion raises, therefore, the question of whether implicatures might be better characterized in terms of joint or we-intentions rather than individual intentions, an issue which has received only passing attention in the pragmatics literature thus far, with the exception of work by Gibbs (1999: 37–38, 2001: 113–114) and Clark (1996, 1997).

The notion of we-intentions remains contentious, however, with a number of scholars arguing that Searle's notion of we-intention is overly internalistic with no clear idea of how they are shared (Fitzpatrick 2003: 60–61; Meijers 2003: 175–179; Velleman 1997: 29). Recent cognitive models of interaction arguably suffer from the same problem, namely that they do not give sufficient detail as to how intentions actually come to be shared (Tomasello and Rakoczy 2003; Tomasello et al. 2005), especially in complex contexts such as communicative interaction.

Another more fundamental problem facing the characterization of implicatures in terms of we-intentions is that it is static notion, and so glosses over the way in which the inferential work underlying interaction is dynamic and emergent. Kidwell and Zimmerman (2006, 2007), for instance, claim that "the capacity for understanding the intentional, goal-directed behavior of others, is fundamentally an interactional process, one that cannot be extricated from the ongoing flow of social activity" (Kidwell and Zimmerman 2007: 592). Through careful analysis of interactions between children showing objects to their mothers or other caregivers, they establish the recurrent exploitation of an "interactional proof procedure" by those participants (594). This interactional proof procedure, which builds on work by Schegloff and Sacks (1973) and Schegloff (1993), is a process whereby "participants make visible their understandings of one another's actions in a sequence of actions…and thereby confirms or disconfirms the other's understanding" (Kidwell and Zimmerman 2007: 594). Such an incremental and sequentially-embedded approach to the establishment of joint attention, and thus indirectly the intentions of others, is not consistent with the assumption that communication is successful once intentions have been "shared," since we can arguably only talk about ongoing *sharing* of intentions, a process which is always contingently relevant on what precedes or follows in interaction.

2.3. Epistemological ambiguity

The epistemological status of the intentions presumed to underlie implicatures remains for the most unstated, although the assumption that hearers ascribe conscious intentions to speakers when drawing implicatures appears common. Recanati (2002: 114–117), for example, claims that implicatures are characterized by the properties of availability, reflexivity, and intentionality. This amounts to the claim that the inferences made about the speaker's underlying intentions are essentially conscious. However, the demand that the inferences and intentions of interactants always be conscious entails "impossible cognitive loads and implausible cognitive processing" (Hample 1992: 317). Kellerman (1992: 293) thus argues that intention should be unlinked from the notion of conscious awareness. Pacherie (2000: 402) and Motley (1986: 3) also claim that many actions do not seem to be preceded by any conscious intention to perform, particularly automatic actions. Yet, it is also apparent through introspection that we can indeed be consciously aware of *a priori* or *post facto* intentions in communicative situations.

It has thus been suggested that we distinguish between "conscious intentions," which vary in their degree of consciousness (Stamp and Knapp 1990), and "cognitive intentions," which do not presuppose conscious awareness (Hample 1992; Heritage 1990/1991). However, since it is only the former that could be argued to guide or motivate speakers in projecting a certain implicature, intention-based theories of implicature in pragmatics are seemingly committed to a conceptualization of speaker intention as conscious mental constructs. But this position becomes problematic when analyzing implicatures in interactional sequences, as apparent from the following example.

(3) (Ronny has called up his friend Sheila on the phone)

1	S:	Hello?
2	R:	'lo Sheila,
3	S:	Yea[:h]
4	R:	[('t's)R]onny.

5	S:	Hi Ronny.
6	R:	Guess what.hh
7	S:	What.
8	R:	.hh My ca:r is sta::lled.
9		(0.2)
10	R:	('n) I'm up here in the Glen?
11	S:	Oh::.
12	R:	hhh
13	R:	A:nd.hh (0.2) I don' know if it's: po:ssible, but
14		.hhh see I haveta open up the ba:nk.hh
15		(0.3)
16	R:	a:t uh: (.) in Brentwood?hh=
17	S:	=Yeah:- en I know you want- (.) en I whoa-
18		(.) en I would, but- except I've gotta leave
19		in about five min(h)utes.=
20	R:	[=Okay then I gotta call somebody else.right away.
21	S:	[(hheh)

(Mandelbaum and Pomerantz 1991: 153–154)

In this excerpt, Ronny starts explaining that his car has stalled and he needs to get somewhere (lines 6–16), but his request appears to have been anticipated by Sheila in line 17, who goes on to imply that she cannot help (lines 17–19). Since neither Ronny's request nor Sheila's refusal are part of what is literally said, they can be understood as being implied in this interaction. The question is, however, to what degree might Ronny and Sheila be consciously aware of their own or the other's intentions? While it appears reasonable to assume that Ronny himself consciously planned to get help (a higher-order intention), an *intention* to which Sheila herself explicitly orients to in line 17 (I know you *want*), it is less obvious that he would necessarily have "consciously attended to all of the many actions, strategies, and communication that he employed as he attempted to get help" (Mandelbaum and Pomerantz 1991: 164). In particular, it is by no means certain that Ronny consciously intended to imply this request. Indeed, the very fact that Sheila anticipated his request in line 17, thereby interactionally achieving the implicature as a conjoint or shared meaning undermines such a position. And while Sheila may have consciously oriented herself to Ronny's *intended* request in line 17, there is little evidence to suggest that the ensuing implicature is consciously intended by Sheila. At best we might assume a level of minimal consciousness (that is, the state of knowing that you know something) as defined by Stamp and Knapp (1990: 283) is involved in the interactional achievement of these implicatures, but there is little evidence to justify holding to the position that any higher level of consciousness is involved.

In summary, then, while the intentions underlying implicatures are canonically understood as an *a priori*, conscious mental state of the speaker about which hearers make inferences, it appears that such a position is not consistent with the ambiguity in the temporal, ontological and epistemological status of such intentions we find when examining actual examples of implicatures in discourse. In particular, such an assumption is found to be inconsistent with the manner in which implicatures are interactionally achieved. In the following section, is thus proposed that what underlies the interactional achievement of implicatures are not inferences about intentions, but rather the general assumption of the "aboutness" or "directedness" of talk (Duranti 2006: 36), to which speakers are held accountable through interaction.

3. The interactional achievement of implicature

The proposed reconceptualization of the interactional achievement of implicatures outlined in this section builds primarily on Arundale's (1999, 2005) Conjoint Co-Constituting Model of Communication. According to this model, communication in general, and thus implicature in particular, should be understand as a type of meaning "that emerges in dynamic interaction as participants produce adjacent utterances and in so doing mutually constrain and reciprocally influence one another's formulating of interpretings" (Arundale 1999: 126). The term "conjointly co-constituted" is employed by Arundale to refer specifically to the conceptualization of communication as an ultimately emergent and non-summative phenomenon. While the Conjoint Co-Constituting Model has implications for the analysis of a large number of pragmatic phenomena, it follows from its key principles that implicatures must be considered from the perspectives of both the provisional meaning that speakers project (Recipient Design Principle), and the hearer's provisional interpretings of the speaker's utterance (Sequential Interpreting Principle), as well as considering how these interpretings are constrained through the adjacent placement of further utterances in conversation (Adjacent Placement Principle) (cf. Haugh 2007: 97). In particular, it follows from the model that interpretings of implicatures by interlocutors are interdependent, since "both persons [are] affording and constraining the other's interpreting and designing" (Arundale 2005: 59). Such a perspective on implicature appears to leave little room for a conceptualization of (speaker) intentions as *a priori* conscious mental constructs, as discussed in the following sections where examples of the two main species of implicature, default and nonce implicatures, are analyzed.

3.1. Default implicatures

Building on Grice's ([1967] 1989, 2001) initial distinction between generalized and particularized implicatures, a number of scholars have argued that default inferences play an important role in communication (Allot 2005; Arundale 1999; Bach 1984, 1995, 1998; Burton-Roberts 2006; Davis 1998, 2003; Haugh 2003; Horn 2004; Jaszczolt 1999, 2005, 2006; Levinson 1995, 2000; Recanati 2002, 2004; Terkourafi 2003, 2005). While there is considerable debate as to whether such default inferences give rise to implicatures, or other species of meanings, most approaches share the basic assumption that associations between utterance types and minimal contexts can give rise, by default, to certain standard meanings.

The notion of default implicature encompasses implicatures that arise in all contexts unless cancelled (generalized implicatures) and those that require a minimal context in which to arise (short-circuited implicatures). Short-circuited implicatures, which encompass cases where the inference underlying the implied meaning is compressed by precedent, but could still be worked out through nonce inference if there were no such precedent (Bach 1998: 713; cf. Horn and Bayer 1984; Morgan 1978; Terkourafi 2003), are one example of pragmatic meaning arising from default inferences. For example, when calling on the phone one might ask "Is John there?" with this being standardly interpreted as a request to speak to John. A minimal context; that is, being on the phone, is all that is required to allow for this implicature to arise. Yet this default inference can be blocked by additional contextual information, for example, that the caller doesn't like John, in which case the caller might be checking if they can talk without being overheard by him. Short-circuited implicatures can be contrasted with the neo-Gricean notion of generalized implicatures (Levinson 1995, 2000), which are activated in all contexts *ceretis paribus*.

Within the framework of the Conjoint Co-Constituting Model of Communication, Arundale (1999) formalizes the insight that default interpretings involve inferential pathways that have been compressed by precedent within the "default interpreting principle:" "If an expectation for default interpreting is currently invoked, and if no conflicting interpreting is present, recipients formulate the presumed interpreting(s) for any current constituent consistent with the expectation" (Arundale 1999: 143). In the case of default implicatures, this principle entails if interlocutors anticipate from the articulation of a particular utterance type in a minimal context that a default inference should be invoked, the implicature that follows from that default inference may be conjointly co-constituted by the interactants, unless there is something that blocks or terminates the process (such as through a non-standard formulation of utterance-type, or through the addition of extra contextual information).

However, if one takes the view that such implicatures arise through default inferences, and consequently rely on expectations shared across speech communities about what such utterance-types standardly imply, there appears to be little room left for the notion of intention, at least in the strong sense of an *a priori* conscious mental construct. Instead, by framing the utterance using expectations about conventions or heuristics the speaker presumes will be salient in interpreting the utterance currently being produced, as well as expectations invoked in interpreting the prior utterance (Arundale 1999: 130), the speaker's utterance only exhibits the more general property of "aboutness" or "directedness." In other words, while the speaker anticipates that a known convention or heuristic will be applicable in deriving an implicature from his or her forthcoming utterance, one cannot maintain the speaker is *intending* to imply per se.

In the following example, an utterance-final disjunctive appears to give rise to a default implicature, namely, the opposite state or condition to that which precedes the disjunctive.[5]

(4) (Chris and Emma have been talking about Emma's acupuncture business)

1	C:	how do you go generally with most of your customers °are
2		they happy or°
3		(0.8)
4	E:	↑YEAH
5	C:	Yeah
6	E:	Yeah I've been getting (0.6) most of my business actually
7		<u>now</u> (0.2) now that it's gaining (0.2) momentum is um word
8		of mouth
9	C:	Mmm
10	E:	From (.) patients telling other patients

11   C: Right (0.5)

In this excerpt, the appearance of "or" at the end of line 2 projects a default implicature something like [or] *not so happy*. Leaving this unsaid may avoid potential impoliteness implications, as well selecting Emma as the next speaker. Preceding Emma's response to the question, there is a slight pause in line 3, which is then followed by a fairly emphatic positive response from Emma in line 4, marked by its rising tone and louder volume. Emma then goes on to describe how her business is growing through word of mouth (lines 6–8, 10), which functions as a warrant for her response. It appears, then, that although the negative state or condition was left unsaid by Chris, Emma displays an interpreting consistent with the co-constitution of this implicature. However, there is no evidence to suggest from this interaction that Emma necessarily perceived Chris as *intending* to imply her business may not be doing so well. If Emma had interpreted Chris' utterance in this manner it would most likely have given rise to implications of impoliteness, and so would have been interactionally marked as such. There is no indication that any such impoliteness implications have arisen in this instance, however, and so Emma does not appear to be holding Chris accountable to this interpreting. In other words, while a default implicature is interactionally achieved here, this arguably follows from a general presumption of the "aboutness" of talk, and understandings of default meanings shared across speech communities, rather than from the ascription of specific intentions to Chris.

Other examples of default implicatures can also be found in Japanese, such as in the following excerpt from a call to a bank by a customer.[6]

(5) (The caller, C, has just gone through the automated answering system and is now speaking to one of the company's representatives, R)

1   C: *Ano:, o-torihiki   hōkokusho tte   arimasu yo ne↓*
                um   Hon-transaction report form Quot exist-Pol FP Tag
                (Um, there's the thing called a transaction report form,
                                                                    right?)
2   R: *Hai.*
                (yes)
3   C: *Sono ken   de   o-ukagai-shi-tai-n-desu   keredomo.=*
                that matter with Hon-ask-do-want-Nomi-Cop(Pol) but
                (It's that I'd like to ask about that matter but.)

4   R: =*Hai. O-uke-itashi-masu*   *node.*
              okay Hon-undertake-do(Hon)-Pol so
              (Okay. It's that I'll handle it so.)
5   C: *Hai. E:to jūichi-gatsu nijūyokka-zuke   de*
              okay uh:m November 24th-   dated   with
6   *ki-te-ru   bun   na-n-desu   keredoMO:*
              come-Te-Prog portion Cop-Nomi-Cop(Pol) but
              (Okay, um, it's that it concerns the one that's arrived which
              is dated November 24th but,)
7   R: *Hai.*
              (Mhm)
8   C: [continues]
(adapted from Yotsukura 2003: 302–303)

The caller's request in lines 1 and 3 for information about a particular transaction that was previously made concludes with an utterance-final *keredomo* (a variant form of *kedo*, 'but'), which not only mitigates this request, making it sound more tentative and thus less demanding, but also is taken up by the other participant as a turn-yielding signal in line 4.[7] The mitigation of the request is accomplished through the implicature that is projected by default through the utterance-final *keredomo* ('but'), namely, that the speaker is leaving other options open to the hearer. The company representative's offer to handle the issue in this line is followed by another utterance-final conjunctive-particle *node* ('so'), which returns the speaking turn back to the caller, as well as implying a request for further details about the transaction that the caller referred to in line 1, or at least allowing interactional space for this further information to be provided. In lines 5–6, the caller begins providing further information and here the utterance-final *keredomo* ('but') once again mitigates the force of the assertion being made through this implicature of leaving options open to the hearer as to how she will respond, as well as signaling its position as a subsequent topic in the caller's continuing narrative. Once again, while it is apparent that these implicatures have arisen, it is less apparent that the interlocutors are being held specifically accountable for those meanings. Instead, the speakers are tacitly understood to "mean" these implicatures only in the sense of being "about" negotiating the interactional flow of this conversation.

It appears, then, that these default implicatures involve appeals to broadly sociocultural defaults (cf. Jaszczolt 2005: 55), and so it is these broad generic expectations (Unger 2006; cf. Green 1995), rather than inferences about intentions per se, that enable the interactional achievement of these implicatures. It is thus suggested that the interactional achievement of such default implicatures does not involve inferences about intentions, but rather a general presumption of the "aboutness" of talk, in conjunction with the invoking of default interpreting norms by interactants.

3.2. Nonce implicatures

The notion of nonce implicatures as unsaid meanings arising from de novo inferences specific contexts draws from Grice's ([1967] 1987) groundbreaking work on particularized (conversational) implicatures. Within the framework of the Conjoint Co-Constituting Model of Communication, Arundale (1999) formalizes the insight that some interpretings involve de novo inferences in the "nonce interpreting principle."

> If an expectation for nonce interpreting is currently invoked, whether because default interpreting was terminated, because no expectation for default interpreting was invoked, or because the expectation for nonce interpreting was invoked explicitly, recipients formulate a particularized interpreting for any constituent consistent with the expectation (Arundale 1999: 143).

In the case of implicature, the nonce interpreting principle indicates that a non-routine inference consistent with the utterance and the context in which it occurs will arise when an expectation for a nonce implicature has been invoked. This expectation can be invoked when (1) a default implicature is blocked, (2) there is no convention associated with that particular utterance (and so there is no expectation for default inferencing invoked), or (3) the nonce implicature is invoked explicitly as the utterance in that context deviates in some way from the expectations of the interactants.

The interactional achievement of nonce implicatures thus involves interlocutors projecting and anticipating particular interpretings, as well retroactively assessing them. In the following example, Emma and Chris are continuing to talk about Emma's acupuncture practice.

(6) (Emma and Chris are talking about how acupuncture draws on the notion of *chi*)

```
1   E:      SO: (0.2) the:y (0.5) they aim to learn to understand it [an:d]=
2   C:                                                              [right]
3   E:      =grow sensitive to it you know, I'm like [(     )]
```

4	C:	[yeah] (0.6) °mmm°
5		(0.2)
6	E:	And the needles happen to be one of the most effective ways
7		to (0.6) manipulate it
8	C:	Yeah
9	E:	Mmmm
10	C:	Can you fix patellar tendonitis? °heh°
11		(1.7)
12	E:	↑maybe ↑yeah
13	C:	Yeah?
14		(0.3)
15	E:	Yeah you got that?
16	C:	I have yeah (0.6) had an operation…

Prior to the projected request in line 10, Emma and Chris have been discussing how acupuncture needles are used to manipulate *chi* and thereby cure various medical conditions. The question in line 10 as to whether Emma could treat a certain condition is thus somewhat abrupt in that there appears to be no discursive work that prepares Emma for this question. While it therefore could be interpreted as a challenge to Emma's medical knowledge, and ability to treat certain conditions, it appears in the utterance sequence following that Emma interprets this as implying a request for help by Chris. There is a relatively long pause (line 11), as Emma apparently considers how to interpret this question, which is then followed by Emma's tentative response in the affirmative that she would be able to treat that condition (line 12). Her interpreting of the question in line 10 as implying a request for specific treatment then becomes apparent in line 15 when Emma specifically asks whether Chris has that particular condition, and Chris confirms that he does in line 16. This interpreting follows Chris expressing interest in her potential ability to treat this condition in line 13, which also projects the question in line 10 as being less of challenge and more an implied request. After discussing Chris' condition for some time, Chris asks later in the conversation for Emma's card, indicating his plan (*intention*) to make an appointment for treatment, which also ratifies the interpreting of line 10 as projecting an implied request.

In interactionally achieving this implicature, then, two points in relation to intention are evident. First, it is not possible to ascertain whether Chris had a particular *a priori* intention in mind when asking about Emma's ability to treat this condition. Even if Chris had been later asked by the researcher what he meant, such *post facto* speculations could only be interpreted in light of how Carl might want to position himself vis-à-vis the researcher and Emma. For example, Chris may not want to admit to the researcher, or even himself, that he had *intended* to challenge Emma's abilities as a medical practitioner in light of the potential impoliteness implications of such a stance. Second, we do not know what particular interpretings Emma might have been entertaining in the long pause that followed Chris' question, or whether she was even considering what "intentions" Chris might have had, as she may simply have been deciding whether she was able to treat the condition.

What is evident, then, is that speculation about the intentions of Chris, or what Emma interpreted as Chris' intentions is analytically unproductive in regards to determining what was implied. Instead, we find it is through Emma's displayed interpreting of Chris' question as an implied request rather than a challenge, and Chris' ratification of this interpreting that this implicature arises. By interpreting the question as an implied request, Emma presumes the question is "about" requesting rather than challenging, and holds Chris accountable to this particular interpreting. In other words, this implicature emerges through the conjoint co-constitution of interpretings in interaction.

In another example involving Kawakami visiting another person's home, Kawakami's request to speak to Naomi is anticipated by the housekeeper, and so the implicature that subsequently arises can be said to have been interactionally achieved.

(7) (Kawakami has arrived at Yabuchi's house. Their housekeeper has just answered the door)

```
1   H:    Hai.
    (Yes)
2   K:    Ano, osoreiri-masu. Ano, watakushi  Nihonjoshidaigaku
          um  feel small-Pol  um   I(Pol)   Japan Women's University
3         no, ano, Fuzokukōkō           no  Kawakami  to
          Gen um auxiliary high school Gen Kawakami Quot
4         mōshi-masu   ga::
          call(Hon)-Pol but
          (Um, excuse me. Um, I'm, um, Kawakami from, um, Japan
          Women's University's, um, high school, but...)
5   H:    Hai. Chotto o-machi-kudasai.
          yes  a little Hon-wait-please
          (Yes, wait a moment please)
```

6   K: *Hai.*
            (Yes)
(adapted from Ikuta 1988: 84–85)

In this excerpt Kawakami first introduces herself after the housekeeper answers the door in lines 2–4. However, before Kawakami goes on to say what her visit is specifically about, the housekeeper anticipates a request to speak to Yabuchi. This anticipation is projected by the housekeeper in asking Kawakami to wait a moment (line 5), presumably while she goes to ask Yabuchi to come to the door. In assuming a request to speak to Yabuchi the housekeeper is thereby interpreting Kawakami's introduction as implying this request, and although it is not necessarily apparent that Kawakami herself is projecting this implicature, she later accepts this interpreting in line 6. The nonce implicature that emerges here is therefore conjointly co-constituted by Kawakami and the housekeeper. In this way, it is apparent that through the housekeeper's projective inferencing in anticipating this request, and Kawakami's retroactive assessing of the inference drawn by the housekeeper, Kawakami is held accountable for this implicature. In other words, the housekeeper infers Kawakami's introduction is "about" requesting, and holds Kawakami accountable for this interpreting.

Nonce implicatures may also arise in interactions where the possibility of something being implied appears to develop over the course of an interaction. In the following example, Kumiko has invited Michiko to play tennis on Saturday. The degree of certainty about whether Michiko is implying a refusal or not varies across the course of this interaction, with the final implicature emerging as an interactional achievement of both interlocutors.

(8) (Kumiko is asking whether Michiko would like to come and play tennis)

1   K: *Deki-tara ne, Mari-san ni   ki-te-itadak-eru   to*
           can-if   Tag Mari-Pol from come-Te-receive(Hon)-Pot Quot
2   *tasukaru-n-da   kedo*
           help-Nomi-Cop but
           (If possible, if you could come that would really help us out but)

3   M: *Aa, tenisu desu   ka. Ii-desu   ne. Tada, chotto doyōbi*
           wa::
           oh tennis Cop(Pol) Q   good-Pol Tag just a little Saturday
           Cont
           (Oh, it's tennis. That sounds nice. But Saturday is a bit…)
4   *Nan-ji   kara desu   ka.*
           what-time from Cop(Pol) Q
           (What time will it start?)
5   K: *Eeto ne, 10 ji   gurai kara yarō   kana to*
           well Tag ten o'clock about from do(Vol) wonder Quot
6   *omot-te-iru-n-da   kedo ne.*
           think-Te-Prog-Nomi-Cop but Tag
           (Um, I think we'll start from around 10 but)
7   M: *Aa, chotto yotei ga   hait-te-shimat-te-iru-n-de,*
           oh a little plan Nom enter-Te-complete-Prog-Nomi-Cop(Te)
           (Oh, unfortunately I already have plans)
8   *Mōshiwake-nai-n-desu   kedo,*
           excuse(Pol)-Neg-Nomi-Cop(Pol) but
           (I am really sorry but)
9   K: *Aa, sō.*
           (Oh really)
10   M: *mata tsugi no   kikai ni demo::*
           again next Gen chance in even
           (if there is a chance next time…)
11   K: *Aa, hontōni.*
           (Oh, really)
12   M: *Ee.*
           (Yes)
13   K: *Sō   ka. Un, wakari-mashi-ta.*
           that way Q yeah understand-Pol-Past
           (Okay, yeah, I understand)

(adapted from Date 2005: 307–308)

In the example above, Michiko has been invited to play tennis by Kumiko. However, Michiko's response in line 3 projects an implied refusal through the utterance-final topic/contrastive particle *wa*. In other words, by saying playing tennis would be good, and then going on to set up "Saturday" as the topic and contrasting it with perhaps some other time, Michiko may provisionally assume that Kumiko would interpret her response in line 3 as implying refusal. However, in line 4 she herself goes on to undermine this provisional assumption in asking what time the tennis game will begin, to which Kumiko responds by giving the time and then leaving it open as to whether Michiko would attend. Michiko once again provisionally projects that Kumiko will interpret her second response in lines 7–8 as implying a refusal, which would most likely also lead to a re-interpreting of Michiko's utterance in line 3 as in fact consistent with this refusal implicature. This provisional implicature is conjointly co-constituted when Kumiko responds with an acknowledgement in line 9 that she has interpreted Michiko's utterances in lines 7–8 as implying she will not play tennis. This implicature is then reinforced through Michiko asking to be included next time in line 10, and Kumiko's subsequent acceptance of that offer in lines 11 and 13.

Looking at this interaction as a whole (including the preceding talk where Kumiko's pre-invitation sequence is extended through a somewhat equivocal response from Michiko), it is apparent it became more likely as the conversation progressed that Michiko would indeed refuse Kumiko's invitation to play tennis. However, it is difficult to pinpoint exactly where in the conversation this refusal was actually implied, and thus the implicature can be regarded as emerging over the course of a number of utterances in the interaction. Thus, while Kumiko presumes Michiko's talk is "about" the invitation, the way in which Kumiko and Michiko engage in both projective inferencing and retroactive assessing in holding Michiko accountable for this implicature, indicates that it is not actually necessary to invoke inferences about Kumiko's intentions in order to explicate this implicature.

3.3. The presumptions of accountability and (emergent) intentionality

In the preceding analysis the role that accountability plays in the interactional achievement of implicatures has been alluded to a number of times. The claim that interlocutors hold themselves and others accountable for meanings that arise from what is said in interaction finds its roots in the work of Garfinkel (1967) and Sacks ([1964]1992: 4–5), but has since been incorporated into conversation analysis (Antaki 1988, 1994; Heritage 1984, 1988) and communication studies (Buttny 1993; Buttny and Morris 2001). It is generally differentiated into two types: normative and moral accountability (Heritage 1988). Normative accountability is essentially "the taken-for-granted level of reasoning through which a running index of action and interaction is created and sustained," while moral accountability involves "the level of overt explanation in which social actors give accounts of what they are doing in terms of reasons, motives or causes"

(Heritage 1988: 128). It is proposed here that interactants are held normatively accountable by others to default or nonce interpretings of what they imply in conversation. In holding interactants accountable for implicatures, norms that are "reflexively constitutive of the activities and unfolding circumstances to which they are applied" are implicitly invoked (Heritage 1984: 109). Building on Garfinkel's (1967) proposals, it is suggested that these norms may be used to account for "perceivedly normal" default interpretings, or to account for "departures" which give rise to nonce interpretings. "The parties to the scene not only maintain and develop the "perceivedly normal" course of the scene by perceiving, judging and acting in accordance with the dictates of the norm, they also use this same norm to notice, interpret and sanction departures from its dictates" (Heritage 1984: 109). In other words, interactants hold each other contingently accountable for default or nonce interpretings of implicatures in interaction through implicit or explicit awareness of these doubly constitutive norms.

It is important to note, however, that related to this presumption of (normative) accountability is yet another presumption which has also been alluded to in the preceding analysis, namely the assumption of the "aboutness" of the cognitive activities underlying talk (Duranti 2006: 36), a notion originally introduced by Brentano in his work on intentionality. This broader concept encompasses:

> the property of 'aboutness' or 'directedness' of the mind...the assumption that all our mental states, such as beliefs, desires, or emotions (hope and fear, love and hate, etc.), as well as perception and behavior, all have an 'object', i.e. they are about something in the world around us (as represented in our knowledge)" (Nuyts 2000: 2–3; cf. Searle 1983: 1).

However, this sense of the "aboutness" or "directionality" of talk "does not presuppose that a well-formed thought precedes action" (Duranti 2006: 33). As Duranti argues: "We might be able to recognize the 'directionality' of particular communicative acts (e.g. through talk and embodiment) without being able to specify whether speakers did or did not have the narrow intention to communicate what is being attributed to them by their listeners" (Duranti 2006: 36). In other words, Duranti (2006) argues, among others (Nuyts 2000: 2; Searle 1983: 1), that the broad notion of intentionality should be clearly differentiated from intention. This presumption is also "persistent" (Mann 2003: 174) in interaction in that it is "neither introduced explicitly nor dismissed upon successful satisfaction." It is important to note, however, the contingently relevant nature of the "aboutness" (or intentionality) of talk, and what is implied by it, as illustrated in the preceding analyses of the interactional emergence of implicatures.

In introducing these underlying presumptions of (normative) accountability and intentionality, then, it has been argued that inferences about intentions are not crucial to the interactional achievement of default and nonce implicatures. As noted by Garfinkel, what speakers want to or intend to mean is not critical in determining what is understood by others. "The big question is not whether actors understand each other or not. The fact is that they do understand each other, that they *will* understand each other, but the catch is that they will understand each other regardless of how they *would* be understood" (Heritage 1984: 119, citing Garfinkel 1952: 367). It is therefore not through inferences about speakers' intentions, but rather through the projecting and retroactive inferencing of interpretings by interlocutors in interaction, and the ever contingent display of interpretings made through responses to previous contributions that implicatures emerge in interaction. Speakers are held accountable to those contingently displayed interpretings of implicature in next turns by others.

4. Is there a place for intention in the interactional achievement of implicature?

In displacing intention with the broader notions of intentionality and (normative) accountability in the interactional achievement of implicatures, the question arises as to what place, if any, remains for the notion of intention in theorizing implicatures. It is suggested in this section that intention may be invoked both explicitly (as a topic of discussion), or more subtly through the discursive practice of what Schegloff (1996) terms "confirming allusions". In both cases, the notion of intention that emerges as salient to the interaction is an emergent *post facto* discursive construct.

A number of scholars have noted in recent times that there are instances where the *intentions* of the speaker themselves can become a topic of discussion, and so becomes a resource for participants through which they can "attend reflexively to the speaker's stake or investment in producing those descriptions" (Edwards and Potter 2005: 246). Recent work by discursive psychologists, for instance, has focused on they ways in which *intentions* are exploited in talk-in-interaction to account for actions (Edwards 2006, 2008). This emergent notion of *intention* is found to be "generally open to formulation, denial, opposition, alternative description, or the partialling-out of intent with regard to specific, formulated components of actions" (Edwards 2006: 44). In other words, through talk about *intentions* or *motives*, speakers can be held *morally* accountable for implicatures (cf. Heritage 1988: 128).

Metapragmatic talk about *intentions* may also involve explicit or implicit reframings of particular implicatures in order to correct "misinterpretations" or accusations of "sneaking through" undesirable meanings. And this talk about the intentions of speakers underlying implicatures can become a very high stakes matter indeed in some cases, as demonstrated by the recent furor over the alleged negative implications for Islam of comments made by the Pope in his speech at the University of Regensburg , or the controversial comments made in a Sydney mosque by Sheik al-Hilali that allegedly implied women who dress inappropriately invite rape (Haugh 2008a).

In the following incident involving discursive dispute of *intentions*, Simon Cowell, a judge on *American Idol* was interpreted as implying disrespect for the victims of the shooting rampage at Virginia Tech. In news media around the world Cowell was reported as being accused of implying disrespect when seen rolling his eyes after a contestant spoke to the victims of the tragedy and their friends and families on the show.

> Simon Cowell is under fire for rolling his eyes at an inappropriate moment. Cowell rolled his eyes and raised his eyebrows as contestant Chris Richardson, of Virginia, followed his performance with a comment about the 32 people killed on the campus by a student: 'My hearts and prayers go out to Virginia Tech. I have a lot of friends over there ... Be strong' (New Zealand Herald, 17 April 2007).

The contestant himself was reported as interpreting Cowell's eye-rolling as implying disrespect towards both himself and the victims, and in this way Cowell was held morally accountable for this implicature.

> Backstage, contestant Chris Richardson was left stunned, calling Simon's reaction "sad and hurtful," a top production source told the DRUDGE REPORT. A backstage source defends Richardson: 'Chris decided earlier in the day that he was going to give a shoutout to his friends...he is hurting' (Drudge Report, 17 April 2007)

In response to this interpreting, the executive producer of *American Idol*, as well as Cowell himself released statements to the media framing this incident as a misunderstanding. For example, in the following excerpt from a subsequent *American Idol* show, Cowell disputed the attribution of the *intention* to imply disrespect.

(9) (Cowell is speaking with the host of *American Idol*)

```
 1   C:    Last night (.) on the sho::w (.) Chris and myself got into
 2         qui'ta <heated debate about> singing through ya ↑no::se↑
 3   H:    Right.
 4   C:    And afte:r (.) a:h I- I finished talking I was talking to Pa:ula
 5         (0.2) and unfortunately I didn't hear Chris (.) mention the
 6         people of Virginia. (0.2) A:h, there was on camera (0.4) I
 7         gave a ↑lo:ok, but in fact I was giving a look to ↑Pa:ula, and
 8         the implication was I was disrespecting (.) the victims. (0.2)
 9         An' I just want to a:bsolutely set the record stra:ight (0.2) I
10         didn't hear what Chris was sa:ying. (0.5) I may not be the
11         nicest person in the ↑world, but I would never ever (.)
           <e:ver>
12         disrespect those families or those victims, and I fe:lt it was
13         important to set the [record straight.
14   H:                        [And [we all know that.
15   A:                             [((cheers))
(American Idol broadcast, 19 April 2007)
```

In this response, Cowell disputes the previously ascribed intention to imply disrespect in a number of ways. He uses stressed intonation on "Paula" in lines 4 and 7 to mark the contrast between his account of his eye-roll as being consistent with what he was saying to Paula (another judge on *American Idol*), and the media's account of it being a reaction to Chris' comments. Cowell also uses repetition and elongation of the word "ever" to distance himself from this interpreting in line 11. His reference to himself as not being widely regarded as "nice" in lines 10–11 projects a possible account for the misunderstanding. And this is further supported by framing his account as a matter of "setting the record straight" in both lines 9 and 13, thereby appealing to norms of "getting a fair hearing." The projected *intention* that Cowell retroactively ascribes to his actions is accepted by the host in line 14 (arguably on behalf of those who were offended), and this is apparently supported by the audience. In this way, talk about *intention* is interactionally achieved in order to dispute a previous interpreting of *intention* that had been attributed to Cowell, indicating that such disputed *intentions* emerge as *post facto* conjoint discursive constructs in interaction.

Another more subtle way in which *intention* may be invoked is through the practice of "confirming allusions" identified by Schegloff (1996). It is argued by Schegloff (1996: 184), drawing from a range of examples, that "agreeing by repeating" may be designed to "indicate a prior orientation to convey" a particular allusion. In the following example, for instance, the customer implies that not all of the plants she bought from Liz died.

(10) (A customer approaches Liz at an outdoor farmers' market)

1	Liz:	Hi::
2	Cust.:	I want to ask you something.
3	Liz:	Sure
4	Cust.:	((pointing to a tray of plants)) I bought three of those uh you
5		know like ( ). One of them died out.
6		((section omitted))
7	Cust.:	Uhh
8	Liz:	It did,
9	Cust.:	Yeah
10	Liz:	The other ones are doing well,
11	Cust.:	The other ones are doing well.
12	Liz:	They were all in the same area,
13	Cust.:	Same thing, yeah.

(Schegloff 1996: 208)

The (default) implicature involved here, namely that the other plants not specified in line 5 by the customer did not die is made explicit through Liz's assertion in line 10 that these other plants are still alive. This is confirmed by the customer in line 11 through an exact repeat. In doing so, Schegloff (1996: 208) claims that the customer not only confirms the allusion, but also confirms it *as* an allusion. In other words, the customer confirms what was implied, as well as indicating her prior orientation (or *intention*) to "convey" this implicature. The customer's "intention" thus emerges as a *post facto* discursive construct through the course of this interaction. In this way, then, it becomes apparent that the invoking of *intention* may become contingently relevant to certain interactions.

A place for intention, albeit somewhat circumscribed, in the explication of implicatures thus remains. In this way, divergences between the traditional explication of implicatures as a matter of hearers attributing intentions to speakers, and the explanation of the interactional achievement of implicature in terms of (normative) accountability and intentionality proposed in this analysis, can be reconciled through the ways in which interactants are held (morally) accountable for implicatures in particular instances.

5. Conclusion

The focus in this chapter has been on reconceptualizing the interactional achievement of implicatures from the perspective of the Conjoint Co-Constituting Model of Communication (Arundale 1999, 2005). Although recent work in language processing argues interactive processes can contribute to overcoming the inherent egocentrism of pragmatic inferencing about "speaker intentions" at the cognitive level (Barr and Keysar 2005), it is suggested here that data which supports the conceptualization of implicatures as an interactional achievement, or non-summative emergence of meaning, requires us to go even further in critically examining the place of intention in theories of implicature, and more broadly communication.

The assumption that the generation of implicatures necessarily involves hearers making inferences about the *a priori* conscious intentions of speakers has been challenged. It has been suggested that when implicatures are examined in actual discourse data, the temporal, ontological and epistemological ambiguity of these intentions becomes apparent. It has thus been proposed that instead of crucially involving these *a priori* speaker intentions, the interactional achievement of implicatures draws on a broader presumption of accountability, namely, that in speaking interlocutors are held accountable for what is implied through the conjoint co-constitution of implicatures. This approach to the interactional achievement of implicatures also draws upon an assumption of the "aboutness" of talk. The place of intention in analyzing implicatures is thus argued to be best reserved only for instances where it is analytically salient.

While cognitive processes undoubtedly underlie the interactional achievement of implicatures, as has been implicitly assumed in this analysis in framing it in terms of such notions as expectations, projecting, retroactive assessing and so on, this does not mean that communication need be reduced to explanations in terms of individual cognition. The notion of "dyadic cognizing," which has been proposed by Arundale and Good (2002) as a means of conceptualizing the cognitive processes underlying communication in a way that is consistent with interactional approach postulated in the Conjoint Co-Constituting Model of Communication, is one possible solution to avoiding this kind of reductionism. A central claim in this approach is that the cognitive processes of participants in conversation are not autonomous but rather are interdependent (Arundale and Good 2002: 127). In other words, rather than being "the summative sequence of individual cognitive activities and/or actions" (Arundale and Good: 124), communication involves what they term "dyadic cognizing." "Each participant's cognitive processes in using language involve concurrent operations temporally extended both forward in time in anticipation or projection, and backwards in time in hindsight or retroactive assessing of what has already transpired. As participants interact, these concurrent cognitive activities become fully interdependent or dyadic" (Arundale and Good 2002: 122).

It is thus argued by Arundale and Good (2002) that while autonomous cognitive processes are clearly involved in human interaction, the traditional "monadic" view of cognition is not able to account for the emergent and non-summative properties of conversation and other forms of talk-in-interaction. It is suggested here that a better understanding of implicatures can be achieved through a shift from traditional explanations of implicatures in terms of concepts such as intention, which are rooted in the minds of individuals, to concepts such as accountability and "aboutness," which are reflective of the fundamentally dyadic nature of cognizing that underlies interaction.

Appendix A: Transcription conventions

The following transcription symbols are utilized:

[ ]	overlapping speech
(0.5)	numbers in brackets indicate pause length
(.)	micropause
:	elongation of vowel or consonant sound
-	word cut-off
.	falling or final intonation
?	rising intonation
,	'continuing' intonation
=	latched utterances

underlining	contrastive stress or emphasis
CAPS	markedly louder
° °	markedly soft
↓ ↑	sharp falling/rising intonation
> <	talk is compressed or rushed
< >	talk is markedly slowed or drawn out
( )	blank space in parentheses indicates uncertainty about the transcription

Appendix B: Morphological gloss symbols

The following abbreviations are used in the morphological gloss:

Cont = contrastive marker

Cop = copula

FP = (utterance) final particle

Gen = genitive

Hon = honorification

Neg = negation

Nom = nominative

Nomi = nominaliser

Past = past tense

Pol = 'polite form'

Prog = progressive

Q = question marker

Quot = quotation

Tag = tag question marker

Te = 'te-form'

Vol = volitional

Notes

1. The view that speaker intentions cannot be dealt with at the utterance level only is further echoed in Fetzer's (2002, 2004) distinction between micro (oriented to single utterances), and macro (oriented to discourse) communicative intentions.
2. A list of standard Conversation Analytic transcription symbols used in the following examples are listed in Appendix A at the end of this chapter.
3. It would be fair to note, however, that Drew (2005: 171) goes on to argue using this example that such cognitive states are "interactionally generated in talk", and so his position is not consistent with the *a priori* view of intention in pragmatics.
4. These are alternatively termed "shared intention(ality)" or "joint intentions."
5. Utterance-final disjunctive particles giving rise to such implicatures were found quite frequently in the corpus of Australian English conversations from which this example is derived.
6. A list of symbols for the morphological gloss is found in Appendix B at the end of this chapter.
7. See Haugh (2008b) for further analyses of (default) implicatures arising from utterance-final conjunctives in Japanese.

References

Allot, Nicholas
    2005   Paul Grice, reasoning and pragmatics. *UCL Working Papers in Linguistics* 17:217-243.
Antaki, Charles
    1988   Explanations, communication and social cognition. In *Analyzing Everyday Explanation: A Casebook of Methods*. Charles Antaki, (ed.), 1-14. London: Sage.
    1994   *Explaining and arguing: the social organization of accounts.* London: Sage.
Antaki, Charles, Michael Billig, Derek Edwards, and Jonathan Potter
    2003   Discourse analysis means doing analysis: a critique of six analytic shortcomings. *Discourse Analysis Online* 1.
Arundale, Robert
    1991   Studies in the way of words: Grice's new directions in conceptualizing meaning in conversational interaction. Paper presented at *International Communication Association*, Chicago, Illinois.

1999   An alternative model and ideology of communication for an alternative to politeness theory. *Pragmatics* 9:119-154.

2005   Pragmatics, conversational implicature, and conversation. In *Handbook of Language and Social Interaction*. Kristine Fitch and Robert Sanders (eds.), 41-63. Mahwah, NJ: Lawrence Erlbaum.

2008   Against (Gricean) intentions at the heart of human interaction. *Intercultural Pragmatics* 5 (2): 231-260.

Arundale, Robert, and David Good

2002   Boundaries and sequences in studying conversation. In *Rethinking Sequentiality. Linguistics Meets Conversational Interaction*. Anita Fetzer and Christine Meierkord (eds.), 121-150. Amsterdam: John Benjamins.

Bach, Kent

1984   Default reasoning: Jumping to conclusions and knowing when to think twice. *Pacific Philosophical Quarterly* 65:37-58.

1987   On communicative intentions: a reply to Recanati. *Mind and Language* 2:141-154.

1995   Standardization and conventionalization. *Linguistics and Philosophy* 18:677-686.

1998   Standardization revisited. In *Pragmatics. Critical Concepts. Volume 4, Presupposition, Implicature and Indirect Speech Acts*. Asa Kasher (ed.), 713-722. London: Routledge.

2006   The top 10 misconceptions about implicature. In *Drawing the Boundaries of Meaning. Neo-Gricean studies in pragmatics and semantics in honor of Laurence R. Horn*. Betty Birner and Gregory Ward (eds.), 21-30. Amsterdam: John Benjamins.

Barr, Dale, and Boaz Keysar

2005   Making sense of how we make sense: the paradox of egocentrism in language use. In *Figurative Language Comprehension. Social and Cultural Influences*. Herbert Colston and Albert Katz (eds.), 21-41. Mahwah, NJ: Lawrence Erlbaum.

2006   Perspective taking and the coordination of meaning in language use. In *Handbook of Psycholinguistics*. Matthew Traxler and Morton Gernsbacher (eds.), 901-938. Amsterdam: Elsevier.

Bezuidenhout, Anne

2002   Generalized conversational implicatures and default pragmatic inferences. In *Meaning and Truth: Investigations in Philosophical Semantics*. Joseph Campbell, Michael O'Rourke and David Shier (eds.), 257-283. New York: Seven Bridges Press.

Bilmes, Jack

1985   "Why that now?" Two kinds of conversational meaning. *Discourse Processes* 8:319-355.

1986   *Discourse and Behavior*. New York: Plenum Press.
1993   Ethnomethodology, culture, and implicature: toward an empirical
           pragmatics. *Pragmatics* 3:387-410.
Bratman, Michael
1987   *Intention, plans, and practical reason*. Cambridge, MA: Harvard
           University Press.
1990   What is intention? In *Intentions in communication*. Philip R Cohen,
           Jerry Morgan and Martha E Pollack (eds.), 15-31. Cambridge, MA:
           MIT Press.
Breheny, Richard
2006   Communication and folk psychology. *Mind and Language* 21:74-
           107.
Burton-Roberts, Noel
2006   Cancellation and intention. *Newcastle Working Papers in Linguistics*
           12-13: 1-12.
Buttny, Richard
1993   *Social Accountability in Communication*. London: Sage.
Buttny, Richard, and G. H. Morris
2001   Accounting. In *The New Handbook of Language and Social
           Psychology*, Peter Robinson and Howard Giles (eds.), 285-301. New
           York: John Wiley & Sons.
Carston, Robyn
1995.   Quantity maxims and generalized implicature. *Lingua* 96:213-244.
1998   Postscript (1995). In *Pragmatics. Critical Concepts. Volume 4,
           Presupposition, Implicature and Indirect Speech Act*. Asa Kasher
           (ed.), 464-479. London: Routledge.
2002   *Thoughts and Utterances. The Pragmatics of Explicit
           Communication*. Oxford: Blackwell.
Clark, Herbert H.
1996   *Using Language*. Cambridge: Cambridge University Press.
1997   Dogmas of understanding. *Discourse Processes* 23:567-598.
Cooren, Francois
2005   The contribution of speech act theory to the analysis of conversation:
           how pre-sequences work. In *Handbook of Language and Social
           Interaction*, eds. Kristine Fitch and Robert Sanders, 21-40. Mahwah:
           Lawrence Erlbaum.
Dascal, Marcelo
2003   *Interpretation and understanding*. Amsterdam: John Benjamins.
Date, Kumiko
2005   An Analysis of Japanese Language Learners' Communication
           Strategies: the Use of Indirect Expressions of Refusal. Unpublished
           PhD thesis, University of Malaya.

Davis, Wayne
   1998   *Implicature. Intention, Convention, and Principle in the Failure of Gricean Theory*. Cambridge: Cambridge University Press.
   2003   *Meaning, Expression, and Thought*. New York: Cambridge University Press.
   2007   How normative is implicature. *Journal of Pragmatics* 39: 1655-1672.
Drew, Paul
   2005   Is *confusion* a state of mind? In *Conversation and Cognition*. Hedwig te Molder and Jonathan Potter (eds.), 161-183. Cambridge: Cambridge University Press.
Duranti, Alessandro
   2006.   The social ontology of intentions. *Discourse Studies* 8: 31-40.
Edwards, Derek
   1997   *Discourse and Cognition*. London: Sage.
Edwards, Derek, and Jonathan Potter
   2005   Discursive psychology, mental states and descriptions. In *Conversation and Cognition*. Hedwig te Molder and Jonathan Potter (eds.), 241-259. Cambridge: Cambridge University Press.
Edwards, Derek
   2006   Discourse, cognition and social practices: the rich surface of language and social interaction. *Discourse Studies* 8:41-49.
   2008   Intentionality and *mens rea* in police interrogations: the production of activities as crimes. *Intercultural Pragmatics* 5 (2): 177-199.
Fetzer, Anita
   2002   Communicative intentions in context. In *Rethinking Sequentiality*. Anita Fetzer and Christiane Meierkord (eds.), 37-69. Amsterdam: John Benjamins.
   2004   *Recontextualizing Context. Grammaticality meets Appropriateness*. Amsterdam: John Benjamins.
Fitzpatrick, Dan
   2003   Searle and collective intentionality. *American Journal of Economics and Sociology* 62:45-66.
Garfinkel, Harold
   1967   *Studies in Ethnomethodology*. Englewood Cliffs, NJ: Prentice-Hall.
Gauker, Christopher
   2001   Situated inference versus conversational implicature. *Nous* 35:163-189.
   2003   *Words Without Meaning*. Cambridge, MA: MIT Press.
Gibbs, Raymond Jr.
   1999   *Intentions in the Experience of Meaning*. Cambridge: Cambridge University Press.

2001   Intentions as emergent products of social interactions. In *Intentions and Intentionality*, Bertram Malle, Louis Moses and Dare Baldwin (eds.), 105-122. Cambridge, MA: MIT Press.

Green, Mitchell. 1995. Quantity, volubility, and some varieties of discourse. *Linguistics and Philosophy* 18:83-112.

Grice, H. Paul
1957   Meaning. *Philosophical Review* 66:377-388.
1989.   Reprint. *Studies in the Way of Words*. Cambridge, MA: Harvard University Press, 1967.
2001   *Aspects of Reason*. Oxford: Clarendon Press.

Hample, Dale
1992   Writing mindlessly. *Communication Monographs* 59:315-323.

Haugh, Michael
2003   Anticipated versus inferred politeness. *Multilingua* 22:397-413.
2007   The co-constitution of politeness implicature in conversation. *Journal of Pragmatics* 39:84-110.
2008a.   Intention and diverging interpretings of implicature in the "uncovered meat" sermon. *Intercultural Pragmatics* 5 (2): 210-229.
2008b.   Utterance-final conjunctive particles and implicature in Japanese conversation. *Pragmatics* 18.

Heritage, John
1984   *Garfinkel and Ethnomethodology*. Cambridge: Polity Press.
1988   Explanations as accounts: a conversation analytic perspective. In *Analysing Everyday Explanation: A Casebook of Methods*. Charles Antaki (ed.), 127-144. London: Sage.
1990/1991 Intention, meaning and strategy: observations on constraints from conversation analysis. *Research on Language and Social Interaction* 24:311-332,

Holtgraves, Thomas
2005   Diverging interpretations associated with the perspectives of the speaker and recipient in conversations. *Journal of Memory and Language* 53: 551-566.

Hopper, Robert
2005   A cognitive agnostic in conversation analysis: when do strategies affect spoken interaction? In *Conversation and Cognition*. Hedwig te Molder and Jonathan Potter (eds.), 134-158. Cambridge: Cambridge University Press.

Horn, Laurence, and Samuel Bayer
1984   Short-circuited implicature: a negative contribution. *Linguistics and Philosophy* 7:397-414.

Horn, Laurence
  2004   Implicature. In *Handbook of Pragmatics*, Laurence Horn and
            Gregory Ward (eds.), 3-28. Oxford: Blackwell.
Ikuta, Shoko
  1988   Strategies of Requesting in Japanese Conversational Discourse. PhD
            diss., Cornell University.
Jaszczolt, Katarzyna M.
  1999   Default semantics, pragmatics, and intentions. In *The
            Semantics/Pragmatics Interface from Different Points of View*. Ken
            Turner(ed.), 199-232. Oxford: Elsevier.
  2005   *Default Semantics. Foundations of a Compositional Theory of Acts
            of Communication*. Oxford: Oxford University Press.
  2006   Meaning merger: pragmatic inference, defaults, and
            compositionality. *Intercultural Pragmatics* 3:195-212.
Jayyusi, Lena
  1993   Premediation and happenstance: the social construction of intention,
            action and knowledge. *Human Studies* 16: 435-454.
Kellerman, Kathy
  1992   Communication: inherently strategic and primarily automatic.
            Communication Monographs 59:288-300.
Keysar, Boaz
  1994   The illusory transparency of intention: linguistic perspective taking
            in text. *Cognitive Psychology* 26:165-208.
  2000   The illusory transparency of intention: does June understand what
            Mark means because he means it? *Discourse Processes* 29:161-172.
  2007   Communication and miscommunication: the role of egocentric
            processes. *Intercultural Pragmatics* 4:71-84.
Keysar, Boaz, Dale Barr, and William Horton
  1998   The egocentric basis of language use: insights from a processing
            approach. *Current Directions in Psychological Science* 7:46-50.
Keysar, Boaz, and Anne Henly
  2002   Speakers' overestimation of their effectiveness. *Psychological
            Sciences* 13:207-212.
Keysar, Boaz, Shuhong Lin, and Dale Barr
  2003   Limits on theory of mind use in adults. *Cognition* 89:25-41.
Kidwell, Mardi, and Don Zimmerman
  2006   "Observability" in the interactions of very young children.
            Communication Monographs 73:1-28.
  2007   Joint attention as action. *Journal of Pragmatics* 39:592-611.
Levinson, Stephen
  1983   *Pragmatics*. Cambridge: Cambridge University Press.

1995   Three levels of meaning. In *Grammar and Meaning. Essays in Honour of Sir John Lyons.* F. Palmer (ed.), 90-115. Cambridge: Cambridge University Press.

2000   *Presumptive Meanings. The Theory of Generalized Conversational Implicature.* Cambridge, MA: MIT Press.

2006a   Cognition at the heart of human interaction. *Discourse Studies* 8: 85-93.

2006b   On the human 'interaction engine'. In *Roots of Human Sociality. Culture, Cognition and Interaction.* Nick J Enfield and Stephen C Levinson (eds.), 39-69. Oxford: Berg.

Locke, Abigail, and Derek Edwards

2003   Bill and Monica: memory, emotion and normativity in Clinton's Grand Jury testimony. *British Journal of Social Psychology* 42: 239-256.

Malle, Bertram, and Joshua Knobe

1997   The folk concept of intentionality. *Journal of Experimental Social Psychology* 33:101-121.

Malle, Bertram

2001   Folk explanations of intentional action. In *Intentions and Intentionality*, eds. Bertram Malle, Louis Moses and Dare Baldwin, 265-286. Cambridge, MA: MIT Press.

2004   *How the Mind Explains Behavior. Folk Explanations, Meaning, and Social Interaction.* Cambridge, MA: MIT Press.

Mandelbaum, Jenny, and Anita Pomerantz

1991   What drives social action? In *Understanding Face-to-Face Interaction. Issues linking goals and discourse.* Karen Tracy (ed.), 151-166. Hillsdale, NJ: Lawrence Erlbaum.

Mann, William

2003   Models of intention in language. In *Perspectives on Dialogue in the New Millennium.* Peter Kuhnlein, Hannes Rieser and Henk Zeevat (eds.),165-178. Amsterdam: John Benjamins.

Marmaridou, Sophia

2000   *Pragmatic Meaning and Cognition.* Amsterdam: John Benjamins.

Meguro, Akiko

1996   Nihongo no danwa ni okeru kansetsuteki kotowari rikai no katei [How native speakers understand indirect refusals in Japanese discourse]. *Toohoku Daigaku Bungakubu Nihongogakka Ronshuu* [Journal of the Department of Japanese, Tohoku University] 6:105-116.

Meijers, Anthony

2003   Can collective intention be individualized? *American Journal of Economics and Sociology* 62:167-183.

Moerman, Michael
   1988   *Talking Culture. Ethnography and Conversation Analysis.*
             Philadelphia: University of Pennsylvania Press.
Morgan, Jerry
   1978   Two types of convention in indirect speech acts. In *Syntax and
             Semantics, Volume 9. Pragmatics.* Peter Cole (ed.), 261-280. New
             York: Academic Press.
Motley, Michael
   1986   Consciousness and intentionality in communication: a preliminary
             model and methodological approaches. *The Western Journal of
             Speech Communication* 50:3-23.
Nuyts, Jan
   2000   Intentionality. In *Handbook of Pragmatics 2000 Installment.* Jef
             Verschueren, Jan-Ola Ostmann, Jan Blommaert and Chris Bulcaen
             (eds.), Amsterdam: John Benjamins.
Pacherie, Elisabeth
   2000   The content of intentions. *Mind and Language* 15:400-432.
Pomerantz, Anita
   1980   Telling my side: "limited access" as a "fishing" device. *Sociological
             Inquiry* 50:186-198.
Potter, Jonathan
   2006   Cognition and conversation. *Discourse Studies* 8:131-140.
Recanati, Francois
   1986   On defining communicative intentions. *Mind and Language* 1:213-
             242.
   2002   Does linguistic communication rest on inference? *Mind and
             Language* 17:105-126.
   2004.   *Literal Meaning.* Cambridge: Cambridge University Press.
Ruhi, Sukriye
   2007   Higher-order intentions and self-politeness in evaluations of
             (im)politeness: the relevance of compliment responses. *Australian
             Journal of Linguistics* 27: 107-145.
Sacks, Harvey
   1992   *Lectures on Conversation.* Oxford: Blackwell.
Sanders, Robert
   2005   Validating 'observations' in discourse studies: a methodological
             reason for attention to cognition. In *Conversation and Cognition.*
             Hedwig te Molder and Jonathan Potter (eds.), 57-78. Cambridge:
             Cambridge University Press.
Saul, Jennifer
   2002   Speaker meaning, what is said, and what is implicated. *Nous* 36:228-
             248.

Sbisà, Marina
1992   Speech acts, effects and responses. In *(On) Searle on Conversation.*
        John Searle, Herman Parret and Jef Verschueren (eds.), 101-111.
        Amsterdam: John Benjamins.
Schegloff, Emanuel
1991   Conversation analysis and socially shared cognition. In *Perspectives*
        on Socially Shared Cognition. Lauren Resnick, John Levine and
        Stephanie Teasley (eds.), 150-171. Washington, D.C.: American
        Psychological Association.
1993   Reflections on quantification in the study of conversation. *Research*
        on Language and Social Interaction 26:99-128.
1996   Confirming allusions: toward an empirical account of action.
        American Journal of Sociology 102:161-216.
Schegloff, Emanuel, and Harvey Sacks
1973   Opening up closings. *Semiotica* 8:289-327.
Searle, John
1969   *Speech acts.* Cambridge: Cambridge University Press.
1975   Indirect speech acts. In *Syntax and Semantics, Volume 3. Speech*
        Acts. P. Cole and J. Morgan (eds.), 59-82. New York: Academic
        Press.
1983   *Intentionality.* Cambridge: Cambridge University Press.
1990   Collective intentions and actions. In *Intentions in communication.*
        Philip Cohen, Jerry Morgan and Martha E. Pollack (eds.), 401-416.
        Cambridge, MA: MIT Press.
2002   *Consciousness and Language.* Cambridge: Cambridge University
        Press.
Sperber, Dan, and Deirdre Wilson
1995   *Relevance. Communication and Cognition.* Oxford: Blackwell.
Stamp, Glen, and Mark Knapp
1990   The construct of intent in interpersonal communication. *Quarterly*
        Journal of Speech 76:282-299.
Taillard, Marie-Odile
2002   Beyond communicative intention. *UCL Working Papers in*
        Linguistics 15:189-206.
Terkourafi, Marina
2003   Generalized and particularized implicatures of linguistic politeness.
        In *Perspectives on Dialogue in the New Millennium.* Peter Kuhnlein,
        Hannes Rieser and Henk Zeevat (eds.), 149-164. Amsterdam: John
        Benjamins.

2005 Pragmatic correlates of frequency of use: the case for a notion of "minimal context". In *Reviewing Linguistic Thought. Converging trends for the 21st century*. Sophia Marmaridou, Kiki Nikiforidou and Eleni Antonopoulou (eds.), 209-233. Berlin/New York: Mouton de Gruyter.

Tomasello, Michael, and Hannes Rakoczy
2003 What makes human cognition unique? From individual to shared to collective intentionality. *Mind and Language* 18:121-147.

Tomasello, Michael, Malinda Carpenter, Josep Call, Tanya Behne, and Henrike Moll
2005 Understanding and sharing intentions: the origins of cultural cognition. *Behavioral and Brain Sciences* 28: 675-735.

Tuomela, Raimo, and Kaarlo Miller
1988 We-intentions. *Philosophical Studies* 53:367-389.

Tuomela, Raimo
2005 We-intentions revisited. *Philosophical Studies* 125:327-369.

Unger, Christoph
2006 *Genre, Relevance and Global Coherence. The Pragmatics of Discourse Type*. Basingstoke, Hampshire: Palgrave Macmillan.

Velleman, David
1997 How to share an intention. *Philosophy and Phenomenological Research* 57: 29-50.

Yotsukura, Lindsay Amthor
2003 *Negotiating Moves: Problem Presentation and Resolution in Japanese Business Discourse*. Amsterdam: Elsevier.

Where is pragmatics in optimality theory?

Henk Zeevat

1. Introduction

This paper deals with the architectural issues of pragmatics within an overall account of natural language in optimality theory (OT). It is argued that pragmatics can be seen as an optimization problem described by its own constraint system that lies outside the constraint system that defines grammar (the production oriented OT models of syntax and phonology). Speaking and hearing both involve grammar and pragmatics, but in different ways. The paper argues against the popular view that grammar and interpretation should be mixed into a symmetric constraint system and connects the proposed architecture with the views that underlie the motor theory of understanding and the mirror neuron theory of the understanding of behavior.

2. The meaning of production

Optimality theory (Prince and Smolensky 1993) can be seen as a modern version of Jakobson's markedness theory. In the very concept of an optimization problem, there is a concept of blocking: some regularity is broken because in the particular case there is a better solution. *Gooses* is ruled out by the "better" *geese* and all the theorist has to do is to explain why *geese* is better. These explanations take the form of a system of constraints, a set of demands on outputs relative to a given input that are linearly ordered by strength. The regularity will always exist, but stronger constraints prevent it from emerging in the particular case.

Though it is not particularly hard to come up with explanations of this kind in phonology and syntax, this is not the business of this paper. For successful treatments of phonology see Prince and Smolensky (1993) and most of the Rutgers Optimality Archive; for syntax an interesting collection is Dekkers et al. (2000). The starting point of this paper is the assumption—really an assumption since many issues remain unresolved—that comprehensive treatments in optimality theory of phonology, the lexicon, and syntax are possible. That successful treatment would allow mapping any meaning to its optimal pronunciation by a function F defined by the constraint system. The inverse of F would be the interpretation function F^{-1} and would deal with semantics and pragmatics. Linguistics would be finished! While this would be nice, there are some problems.

A first problem is the role of the context. A proposition like *Tim is happy* can be expressed in a number of ways depending on the context.

(1)   Yes.
    He is.
    She is.
    He is however.
    He is too.
    He is happy.
    ...and happy.
    Tim is happy.

The variation depends on the conversational setting (e.g. did the interlocutor ask: Is Tim happy?) on the degree of activation of Tim in the context and on the degree of activation of the predicate *be happy* and on the presence of reasons for thinking he might not be happy (*however*) or other happy people (*too*). Intonational variation is not included in the example, but would give rise to a whole range of further variation.

With this addition, F^{-1} will assign sets of pairs of contexts and meanings and since the interpreter presumably knows the context, the possible meanings of the utterance u in the context c can be defined as the set $\{m: u$ is optimal for m in $c\}$.

Even with this addition, there are problems. First of all, there exists semantic blocking next to the production blocking discussed before. These are examples like:

(2)   a. Katja and Henk$_i$ were surprised that the journal rejected each other's$_i$ papers.
    b. Katja and Henk$_i$ were surprised that the editors$_j$ rejected each other's$_{j/i}$ papers.

(3)   Poor Jones kicked the bucket. (non idiomatic)

(4)   Jones sat down on the bank (financial institution)

(5) John has three cows. (the at least 3 reading)

(6) How late is it?

In (2b) *each other* cannot take the antecedent it takes in 2a, because the antecedent *the editors* is strongly preferred. In appropriate contexts, 3 will be interpreted in the idiomatic way with blocking of the literal context (this example is slightly problematic: it has been argued that this sort of idiom always evokes the literal meaning. The literal meaning is there in one sense, but it is not there as the intended meaning.) In 4 the difficulties of sitting down on financial institutions repress that meaning of *bank*. Example (6) is related to syntactic blocking. It seems that because this should have been said as *what time is it?*, it cannot have the meaning that according to compositional semantics it should have.

The problem with the interpretation function F^{-1} is that it is not given as an optimization problem and that thereby it is unable to implement the idea that there are better interpretations that block the blocked interpretations.

Hendriks and de Hoop (2001) shows that it is fruitful to think of interpretation as an optimization problem and that interestingly it needs to take account of all the formal factors involved in production: syntax, lexicon, intonation and context. It is hard to see how traditional accounts of semantics could deal with the problems that are dealt with in the paper.

A further problem comes from the proper application area of pragmatics. Pragmatics proper needs to deal with problems such as interpretation preferences given the context, stereotypicality effects, resolution of anaphora and presupposition and implicatures arising from relevance and other sources. These factors cannot be dealt with in production. The pragmatically dispreferred interpretations can still be possible inputs in the context and so would be mapped by F to the utterance. The utterance however will not have the interpretation in the given context, though it may well have it in another context.

Take for example the familiar defaults about presupposition resolution and accommodation. The production constraint on the use of a trigger like *regret* should be limited to the requirement that the local context entails the complement, in (7) that Tim married Mary. The two interpretations (7b) and (7c) therefore seem to be alright if the context does not have the information that Tim married Mary. And in fact, in a context in which it is entailed that the speaker does not know that Tim married Mary (e.g.

because it contains information that he did not) (7b) is the right interpretation while in contexts in which the speaker could know this, (7c) is the best interpretation. Restrictions on production alone cannot give this preference.

(7)   a. If Tim regrets marrying Mary, I would be surprised.
      b. If Tim has married Mary and regrets it, I would be surprised.
      c. Tim has married Mary and if he regrets it, I would be surprised.

The considerations above lead into a confusing situation. Suppose, following Hendriks and de Hoop (2001), that a constraint system SEM can be developed that defines a function G analogous to the function F that can be defined from the production OT constraint system, but this time from utterances in a context to interpretations. F and G must be related but how? There are two implications that seem plausible ways of relating the functions:

$$u \in F(m,c) \rightarrow m \in G(u,c)$$
$$m \in G(u,c) \rightarrow u \in F(m,c)$$

The presupposition example gives a counterexample to the first implication. But fortunately counterexamples of this kind can be ruled out by restricting
F to G:

$$F'(m,c) = \{u \in F(m,c): m \in G(u,c)\}$$

(This seems a vindication of the pruning strategy proposed by Blutner 2001. It will however turn out in section 3 that pruning as practiced also cuts out perfectly healthy branches.)

In this restriction, the speaker is pictured as somebody who steps into the hearer's shoes and disallows possible utterances that will be misunderstood.

How about the second implication? Doubt is cast on the principle by the familiar fact that language users understand many utterances they would never produce, either at any stage of acquisition or afterwards. This can in the context of OT be dealt with: an OT constraint system does not just define the best utterance, but it also induces an ordering over the set of all possible utterances. All that needs to be done is to exploit this ordering in the following way.

$m \in G'(u,c)$ iff there is no $m' \in G(u,c)$ such that u is better for m' than for m in the production system.

In this definition it is the hearer who is charitable: no matter the quaintness of the speaker's way of expressing herself, the hearer makes the best of it.

G' and F' are improved versions of F and G and it can be shown that $F'^{-1} \subseteq G'$.

Can F' and G' be implemented by constraint systems? There are two proposals here. The first is known as bidirectional optimality theory and is quite popular, with Boersma, Blutner, De Hoop, Hendriks, Spenader, Bouma and De Swart coming out in favor of it. The idea is to have a single constraint system comprising the production systems and semantic constraints that computes both F' and G' by computing the best utterance for a meaning and by computing the best meaning for an utterance. There are two flavors of this idea that are often not well distinguished. In the first flavor, the computation of the meanings or forms is conditioned by the computation in the other direction. In the strong version, m is optimal for u if m wins the competition for u and u wins the competition for m. It follows that m can only be optimal for u if u is optimal for m. (I will not discuss the weak version here.) In the second flavor, the constraint system itself is assumed to be symmetric: it has the property that u wins for m iff m wins for u.

I will come back to these proposals in section 4 after here presenting my own proposal, the motor theory of language understanding. In this theory, production OT is adopted without any changes. It is after all possible to give illuminating and correct descriptions of phonology and syntax by production constraint systems. These production systems -taken in conjunction- also define grammar: the relation between form and meaning in a context[1]. On top of that there is pragmatics: a separate optimization problem in which it is decided which of the grammatical meanings of a given form is to be preferred.

The production systems are learned by the users of languages and have emerged from language evolution. OT gives an account of the plasticity in learning: next to the lexicon, it is only the ordering of the constraints, and so restricts the learning problem and the typological possibilities. The pragmatic system in contrast does not need to be learnt and has been accepted as is: it is the system with which humans and their ancestors make sense of intentional communicative behavior directed at themselves and, as such, it predates language. Section 2 gives a brief introduction to this style of OT pragmatics.

A similar proposal has been made in the context of OT phonology by Hale and Reiss (1998) and it has a problem that my proposal has to face as well. There is no optimality theoretic account of how the production system is inverted. And this has repercussions for popular views of OT learning such as Boersma (1998) or Tesar and Smolensky (1998). My original view was that this is just a question of computation and a solved one at that. Frank and Satta (1998) and Karttunen (1998) get very close^2 to showing that OT phonology can be inverted by compiling the finite state transducers. One can similarly use one of the popular stochastic parsers (Manning and Schuetze 1999) to come up with reasonable candidate interpretations that can be checked against production. These technologies are there and can be used without any problem for OT learning. The only thing that matters is whether one's own production for the understanding can match the utterance or not.

It is however not at all obvious that not having an OT account of the inversion of production and having to rely on what appear to be engineering approaches in this area is such a problem. The first consideration is that in natural language parsing and even more strongly in speech perception one should have very serious doubts whether rule-based approaches are practically feasible at all and if they were whether they are as learnable as the stochastic approaches that have become standard in these areas.

What is more, stochastic recognizers can in principle be made to have the same bias as the pragmatic constraints I will discuss later: it is just a question of choosing the appropriate stochastic variable they should minimize. And getting them to be like that merely increases their similarity to what we seem to understand about the working of our brains: contextual activation and associative processes are central in their operation and central to perception. For the pragmatics of section 2, a case can be made that it is nothing else but context-driven perceptual bias in the distal perception of the speaker intention.

So there is a good case for not taking these stochastic parsers as just engineering tools that have to be used because there are no proper tools yet or not enough of them, but as approximate models of what goes on in human speech and language recognition. The OT production models and

OT pragmatics can help in making them better approximations of what goes on in human language perception.

The second consideration comes from the discovery of mirror neurons. This research originally showed that in the F5 pre-motor cortex of rhesus monkeys there is a class of neurons that fire both when an action like grabbing an object or tearing it up is planned but also when the monkey perceives another organism do the same thing. What it shows in short is that the part of the brain that is responsible for planning the action plays a presumably important role in the perception of the same action. This role is presumably important, because otherwise the brain would not have evolved to create this activation pattern in perception. Meanwhile mirror neurons have been discovered in many other parts of the brain and in many other species, including humans. Rizzolatti and Arbib (1998) speculate that F5 is the precursor of Broca's area. Mirror neurons open the door for accounts of understanding behavior of other organisms by reconstructing it as behavior of themselves.

In speech recognition, Alvin Liberman (e.g., Liberman and Mattingly 1985) is the author of the motor theory of speech perception. There is some debate about this theory in speech perception but Liberman's theory is rather minimal. It holds first of all that perception of speech is distal perception of the articulatory gestures that the speaker makes in producing the speech. Liberman came to this theory because he thought these were the real invariants in speech: the acoustic signal is too much spoilt by biological differences between people and by coarticulation. The debate on the motor theory of speech perception is about the proper invariants in the speech signal and there are arguments for thinking that Liberman was wrong or partly wrong. The second part of the theory was largely speculative when Liberman formulated it but is now abundantly confirmed. It is that the articulatory parts of the brain play a prominent role in speech perception. I am hardly a speech technologist, but it would seem to me that if it is correct to identify phonemes as bundles of articulatory features and if one tries to take the standard hidden Markov model speech perception seriously from a cognitive perspective, then what happens in that approach is distal perception of phonemes, by trying to maximize the probability of a certain phoneme as causing the signal multiplied with the probability of the phoneme in the context. The discussion about invariants is hardly relevant for taking Liberman's theory seriously in this respect. An important element of Liberman's argument is *parity*. The idea is that in a communication system it should be possible to explain how the sender and the receiver can converge on the same signal. In the case of speech, for the speaker the signal is a complex of articulatory gestures, for the hearer an acoustic signal. There must be some point at which the speaker and the hearer agree on the identity of the signal. Liberman's proposal is that the hearer reaches identity by recognizing the articulatory gestures. The truth may be more in the middle, but reconstruction of the articulatory gestures is still part of reaching parity.

Another forerunner of these thoughts (around the same time as Liberman) is Grice. In Grice (1957), non-natural meaning is defined in terms of intention recognition. It follows from the Gricean definition that communication fails if the hearer does not recognize the intention. But it seems a mystery what intention recognition is. I used to be quite puzzled by what this could be. It now seems quite obvious: the hearer should reconstruct the whole action of the speaker in producing the utterance as a possible action of one's own. If the reconstruction is successful, it is recognition of the speaker's intention and a side effect would be the reconstruction of all the judgments that underlie the various choices the speaker made in producing the utterance.

The absence of an OT-based account for the reversal of production OT is therefore not a problem but an asset. It allows general perceptual mechanisms to take over and the motor theory of understanding makes it understandable how this could work: the perceptual mechanisms distinguish different possible states of the production system. It is therefore the simplest and most natural explanation of parity.

3.   The OT pragmatic system

This section gives a brief overview of the optimality theoretic pragmatic system focusing on production OT. More elaborate treatments are described in Zeevat (2001, 2007b, and 2007c), while Zeevat (2007a) discusses the consequences for presupposition projection.

The pragmatic constraints can be seen as a definition of what is marked in interpretation. First of all, the interpretation of the utterance must be an explanation of the utterance. The speaker must understand from the interpretation why it was made and why it led to the particular signal that was produced. The interpreter can judge the quality of the explanation in this respect because the interpreter is also a speaker and knows the context and can consequently simulate the production. In the theory of this paper, production OT defines the rules of the process, starting at the maximally abstract level (the intention of the speaker) and going down to the level of the speech. Marked according to this constraint are any deviations from what is overtly given in the utterance. One of the predictions from this principle is therefore that non-literal interpretations only occur if literal interpretations do not succeed.

The second constraint is plausibility. It should not be the case that there is an equally good interpretation that is more plausible. In this notion of plausible there should be several layers. One level is purely linguistic: if there are ambiguities, the most likely interpretation should be chosen based on probabilities given in language use. The other side of plausibility is the probability of the message in the context. This can go from the context ruling it out entirely to its being surprising in the context and from there to it being expected or fully known. The last cases are the unmarked ones. The most unmarked is the most expected. FAITH however does not allow interpretations of an utterance in which there is no point to the utterance.

The third constraint *NEW enforces conservatism with respect to the context. If referents have to be assumed in the interpretation, one should always prefer the referents with the highest activation level possible. Fully new referents come last. Given that interpretations have referents of various kinds (objects, moments of times, events and states and maybe even topics) this forces maximization of coreference. The unmarked case is that the utterance stays with the entities and topics that were under discussion. This fits with the idea that most unmarked rhetorical relation is the *restatement* relation as shown by Jasinskaja (2007).

The last constraint, relevance, prefers interpretations which help to achieve current goals of the conversation or which settle questions that have been activated. From the perspective of this constraint what is unmarked makes sense with respect to the goals of the conversation. Digressions and attempts to address a new topic are special.

1. FAITH: there is no interpretation for the utterance for which the hearer –putting herself in the position of the speaker–could have produced an utterance that is closer to the given utterance.

2. PLAUSIBLE: maximize plausibility (an interpretation is bad if there is a more plausible interpretation that is otherwise equally good)

3. *NEW: old referents are preferred over connected referents which are preferred over new referents.

4. RELEVANCE: let the interpretation decide any of the activated questions it seems to address.

The constraints must apply in this order. FAITH should be able to override any of the concerns of the other constraints. PLAUSIBLE is–in its form of a consistency checker–a well-known constraint on pronoun resolution and on relevance related implicatures, including presupposition accommodation. The placement of *NEW over RELEVANCE can be argued from presupposition accommodation: *NEW says that presupposition resolution is always preferred even if a more relevant reading can be reached by accommodating as in (8).

(8)   If John is rich, his wife must be happy.
      If John is married, his wife must be happy.

While it would definitely help in settling the question whether John is married (achieved by a global accommodation), as in (8a), resolution is possible in (8b) which makes accommodation impossible.

The system follows the architecture of relevance theoretic pragmatics: it can be interpreted as adding information to underspecified interpretations. The speaker however is monitoring the pragmatic effects her utterance may have and will change the utterance if unwanted effects are predicted. The system can be used to interpret non-linguistic communication without any essential change.

As an example, consider the following situation. John stands by the road waving with his jacket at me. I should be asking myself first when I would be standing by the road waving my jacket at someone. This is FAITH: it requires me to have an explanation. The possible answers should be weighed by plausibility and better answers should be selected over less plausible ones (PLAUSIBLE). I should then be wondering about the new elements in my explanation, can they be eliminated, can I connect them to already known things (*NEW)? And if there are activated questions, can John be settling them by his waving (RELEVANCE)? If one supposes that we were looking for a lost cow, the proper explanation may be that John has seen it and that he is indicating where it is. The jacket waving then means: I found the cow. Here it is!

The first three constraints can also be understood as an OT account of explanation in general. FAITH should then be reinterpreted as the check that what is offered as an explanation would in fact have caused the explanandum. The other two constraints then maximize the plausibility of the explanation and minimize the new assumptions occurring in it.

This paper does not have space to apply the constraint system to the whole field of pragmatics and–anyway–the relevant areas are covered in the papers referred to above. Only some examples will be given therefore. But the constraint system has the ambition to cover the whole of pragmatics and this appears to work: the same principles play a role in rhetorical structure, in presupposition resolution and accommodation, in implicature projection, and in pronoun resolution. This cannot be avoided within OT methodology: a constraint cannot just be switched off when one goes to a different area of pragmatics. The constraints are generalizations from assumptions in the standard Heim/Van der Sandt theory of presupposition projection (Heim 1983; Van der Sandt 1992) and it was quite surprising that they turn out to have an explanatory value on rhetorical structure, pronoun resolution and implicatures, areas that these researchers on presupposition did not take into account in these papers.

3.1. Example 1: the pronoun "she"

The production system allows the pronoun only for singular referents with female agreement and a high level of activation. This high activation level can be due to the non-linguistic context, but more commonly to mention in the pivot or in the current sentence. The production system also prefers "she" if the conditions on its use are fulfilled. Production monitoring for the reference feature prevents "she" from being used for Mary after examples like:

(9)   Mary and Jane went to the cinema.

FAITH guarantees that the possible referents are highly activated, otherwise the use of "she" cannot be explained.
PLAUSIBLE can rule out antecedents.
*NEW's only role with respect to the pronoun is to guarantee that there is not a higher activated suitable antecedent.
RELEVANT can decide between equally ranked antecedents.

3.2.   Example 2: "a new rucksack"

(10) Bill went to Spain. He bought a new rucksack.

The production system prefers the indefinite marker on the rucksack only if the conditions for definite marking are not fulfilled. One of these is that there is no unique description available and in particular that "new rucksack" is not a unique description of the referent. It allows "a new rucksack" only if the referent is a new rucksack.

FAITH reconstructs these considerations. The referent must be new to the context (otherwise there would be a definite alternative) and "rucksack" is not a unique description. Together this forces the construction of a new discourse referent for the rucksack.

*NEW prefers a connected discourse referent to a fully new one. Since rucksacks play an enabling role in traveling, the inference is reached that this is the rucksack Bill used in traveling to Spain. Since the rucksack cannot be used before it is acquired, this also forces the buying to be before the traveling. (The reasoning around the buying event is similar, and it is hard to say in which order the inference is actually reached).

RELEVANT is responsible for answering the question why Bill bought the rucksack and so for placing the buying in the preparatory phase of going to Spain. And for exhaustivity implicatures: he bought nothing else of the same significance in his preparation.

3.3.   Example 3: the particle and sentential conjunction "and"

(11) Bill left. And Martha followed him.

As argued in Jasinskaja and Zeevat (2007), "and" is a strongly grammaticalized additive marker ("also" is less grammaticalized, "in addition to that" not at all). It imposes on its use in a clause that there is another clause belonging to the same topic which is distinct from it. The grammaticalization makes it possible that this is not really so, but only according to a prominent view in the context. This covers the cases in which it is allowed. Monitoring must be assumed for the "additive" feature: same topic, different element and this would entail that "and" or a replacement is obligatory in certain cases.

FAITH reconstructs these considerations. In particular it forces the identification of the shared topic and of the other element, the assumption of distinctness between the identified element and the current one or identifies the view under which they are distinct.

"And" belongs to the particles which switch off *NEW through FAITH. But *NEW still forces a preference for the most activated antecedent for "and" and can thus be made responsible for the formation of the version of "and" in which it is a sentential conjunction.

4. Speaking and hearing

A speaker is also a hearer and as such can bring in expectations about how she is going to be interpreted into the decisions about the form of what she is going to say. There is a proviso here: the formal possibilities for going to a different formulation should be there and sometimes they are not.

(12) Welches Maedchen mag Peter?
  Which girl likes Peter/does Peter like?

In (12), the word order dimension has been exploited to mark the sentence as a question and to mark the wh-element by fronting the wh-NP. That means that unlike in (13), it cannot be used again for marking the subject. As a result, the sentence is ambiguous: Peter can be the subject or the object.

(13) Peter mag Maria.
  Peter likes Maria.

In (13), there is a strong preference for canonical subject object order which marks Peter as the subject (case marking and agreement do not mark it in this case.) This means that the word order dimension is again fully used and not available for a third task to which it may be set: the marking of contrastive topics, as in 14 ("er" is the nominative of the male pronoun).

(14) Maria mag er. (Aber nicht die Christina).
  He likes MARIA (but not Christina).

The strong preference in (13) can be explained as the speaker monitoring the hearer: the speaker checks whether the hearer can find out who likes who and if necessary adjusts the word order to subject before object. In

(14), this is taken care of by the case marking and no adjustment of word order is necessary.

A formulation in which monitoring is absolute (the speaker refuses a formulation unless it is guaranteed that the hearer will understand it correctly, i.e. the version F' of F from section 1) will lead to problems. (12) will be disallowed unless *welches Maedchen* is the subject, (13) will be disallowed when *Maria* is the contrastive topic (assuming monitoring for the expression of the contrastive topic).

Monitoring seems to happen in this moderate way and with priority for certain features over others. It is about the phenomenon of optional marking and the explanation of why certain optional marking strategies are obligatory when they occur in a larger text.

$F \cap G^{-1}$ as proposed in section 1 is therefore too strong. It does not need to be so strong either in the current perspective. Interpretation outperforms speaking. If one can prevent confusion by marking one should do so. But if marking is impossible or hard, it is still more likely than not that the understanding will nevertheless be correct.

In interpretation there is no corresponding monitoring in the motor view: understanding is identical with finding the least marked reconstruction of speaking. Understanding is identical to what would be monitoring.

5. Parity

Smolensky (1996) is one of the earliest applications[3] of bidirectional or symmetric OT and it attempts an explanation of why young children can understand things they cannot produce yet. To borrow one (made up) example, the child will produce the name Kate as /ta/ but will understand /kaet/ as Kate[4] and /ta/ as "ta". The imperfect production is explained by the high ranking of markedness constraints with respect to faithfulness constraints. This will produce low marked forms in production, but since markedness constraints do not play a role in understanding, faithfulness will produce understandings that are similar to the adult case. Language learning then demotes the markedness constraints until the point where a symmetric system results.

Hale and Reiss (1998) is an attack on the whole line of reasoning of Smolensky's paper, but especially on the idea that production should be inverted and that symmetry of the constraint system will eventually result.

To show this, a simple counterexample is given: the two German words *Rat* and *Rad* that share their pronunciation /rat/ due to FINAL DEVOICING, a constraint outranking FAITH (VOICE). In the production direction this gives a correct description of the phenomenon, but in the interpretation direction, the interpretation *Rad* for /rat/ will incur a FAITH (VOICE) error that the interpretation Rad does not get. The adult constraint system is therefore not symmetric and no degree of further learning will make it so. And one cannot prune it into symmetry by means of bidirectionality, since then *Rad* with its pronunciation /rat/ just disappears. Hale and Reiss have an important conceptual point. No ambiguity should ever be resolved by the inverse competition. This is not how ambiguities are resolved; they are solved by semantic considerations and not by phonology or syntax. Inverting production competitions will however just open the possibility that they will.

The Rat/Rad problem also establishes two negative conclusions about parity. Just running the inverse competition with the production constraints does not establish parity unless the system is symmetric. And bidirectional pruning also does not give parity on a correct but asymmetric system; it makes a correct production system incorrect.

The solution proposed by Boersma (2001) is the only proper one: add sufficiently many semantic constraints to make the constraint system symmetric without destroying the behavior in the production direction.

Unfortunately, there is no theorem that says that this can always be done. Boersma's example is Dutch phonology taken as a finite relation between lexical items and surface forms. And for finite relations, one can prove that semantics can be done by a semantic OT constraint system such as the one proposed by Boersma. From this constraint system, one can then construct the symmetric system. But can this be generalized to the infinite case? And I am not sure either that Dutch phonology should be seen as a finite problem. It seems that a proper account of Dutch phonology should also predict that the fantasy pronunciation /tat/ is ambiguous between the fantasy words *tat* and *tad*.

So, the problem of there being a symmetric system incorporating any given correct production system is fully open. And full symmetry has the problems noted in section three, if the pragmatics of section two is integrated.

While there is no theorem, there is also no counterexample to the claim that any correct production system can be inverted by another constraint system. But I have an argument against the claim that there always is.

Languages like Dutch, English or German have acquired a vast functional inventory by a process that is called grammatical recruitment of originally lexical words. This would happen in the cultural evolution processes that shaped these languages and would require a functional explanation. The simplest explanation is that recruitment happens to improve the chance that one is understood properly: partial recruitment leads to improved understanding which leads to increased reproduction of the recruited item in its new role. Now how could this ever happen if the production system has a perfect OT inverse? Very much the same point can be made from the study of dialogue as in Clark (1996). One of the clearest findings in that work is that there are powerful feedback mechanisms to monitor proper understanding and supply feedback. This would be fully unnecessary with full symmetry.

It would therefore seem that the motor theory of understanding is so far the only of the views considered in this paper that can account for parity in the adult system.

Notes

1. One can conceptually add pragmatics to grammar, as in the definition of F', but then grammar is no longer given by an OT production system alone. I prefer to call grammar grammar, and pragmatics pragmatics, and to refer to the integrated notion as speaking. This is not speaking as it is practiced around us but an idealization, the behavior of undisturbed competent speakers.
2. The construction needs an upper bound on the number of errors that a given constraint can assign to a candidate. This is not pure OT, but it gets very close.
3. The idea comes from OT learning: high markedness constraints and low faithfulness constraints produce a robust parser in the opposite direction and a learning mechanism based on it can demote the markedness constraints.
4. This is demonstrated presumably, by semantic understanding of the name: the child will look at Kate. I am forced to claim that stochastic understanding is in place already. The sound /kaet/ is associated with *Kate*. The imperfect rendering of *Kate* as /ta/ is a learning datum which will demote the markedness constraints. At this stage, classification by production cannot play a serious role yet.

References

Blutner, Reinhard
    2000   Some aspects of optimality in natural language interpretation. *Journal of Semantics* 17: 189–216.

Boersma, Paul
    1998   Functional phonology: Formalizing the interactions between articulatory and perceptual drives. Ph.D. diss., University of Amsterdam.
    2001   Phonology-semantics interaction in OT, and its acquisition. In *Papers in Experimental and Theoretical Linguistics, Vol. 6*, Robert Kirchner, Wolf Wikeley, and Joe Pater (eds.), 24–35. University of Alberta.

Clark, Herbert H.
    1996   *Using Language.* Cambridge: Cambridge University Press.

Dekkers, Joost, Frank van der Leeuw, and Jeroen van de Weijer, J. (eds.)
    2000   *Optimality Theory: Phonology, Syntax and Acquisition.* Oxford: Oxford University Press.

Frank, Robert, and Giorgio Satta
    1998   Optimality theory and the generative complexity of constraint violability. *Computational Linguistics* 24:307–315.

Grice, H. Paul
    1957   Meaning. *Philosophical Review* 67:377–388.

Hale, Mark, and Charles Reiss
    1998   Formal and empirical arguments concerning phonological acquisition. *Linguistic Inquiry* 29:656–683.

Heim, Irene
    1983   On the projection problem for presuppositions. In *Second Annual West Coast Conference on Formal Linguistics*, Michael Barlow, Dan Flickinger, and Michael Westcoat (eds.), 114–126. Stanford, CA: CSLI Publications.

Hendriks, Petra, and Helen de Hoop
    2001   Optimality theoretic semantics. *Linguistics and Philosophy* 24:1–32.

Jasinskaja, Katarina
    2007   Pragmatics and Prosody of Implicit Discourse Relations: The Case of Restatement. Ph.D.diss., University of Tuebingen.

Jasinskaja, Katarina, and Henk Zeevat
    2007   "And" as an additive particle. In *Language, Representation and Reasoning. Memorial Volume for Isabel Gomez Txurruka*, Mixel Aurarnague, Kepa Korta, and Jesus M. Larrazabal (eds.). UPVEHU, Bilbao.

Karttunen, Lauri
 1998   The proper treatment of optimality in computational phonology. In
         *FSMNLP '98: Proceedings of the International Workshop on Finite
         State Methods in Natural Language Processing.*
Liberman, Alvin M. and Ignatius G. Mattingly
 1985   The motor theory of speech perception revised. *Cognition* 21:1–36.
Manning, Christopher, and Heinrik Schuetze
 1999   *Foundations of Statistical Natural Language Processing.*
         Cambridge, MA: MIT Press.
Prince, Alan, and Paul Smolensky
 1993   Optimality theory: Constraint interaction in generative grammar.
         Technical report, Rutgers University Center for Cognitive Science
         Technical Report 2.
Rizzolatti, Giacomo, and Michael A. Arbib
 1998   Language within our grasp. *Trends in Neurosciences* 21:188–194.
Smolensky, Paul
 1996   On the comprehension/production dilemma in child language.
         Linguistic Inquiry 27:720–731.
Tesar, Bruce, and Paul Smolensky
 1998   Learnability in optimality theory. *Linguistic Inquiry* 29:229–268.
Van der Sandt, Rob
 1992   Presupposition projection as anaphora resolution. *Journal of
         Semantics* 9:333–377.
Zeevat, Henk
 2001   The asymmetry of optimality theoretic syntax and semantics.
         Journal of Semantics 17(3):243–262.
 2007a   A full solution to the projection problem. Unpublished manuscript.
         University of Amsterdam.
 2007b   Optimal interpretation as an alternative to Gricean pragmatics.
         Unpublished manuscript. University of Amsterdam.
 2007c   Optimal interpretation for rhetorical relations. Unpublished
         manuscript. University of Amsterdam.

Intention, common ground, and the availability of semantic content: a relevance-theoretic perspective

Stavros Assimakopoulos

1. Decoding and inference in linguistic communication

By its very definition, communication involves the transmission of information from one agent to another; if the communicator had no means of making some piece of information available to an audience, there would *a priori* be no possibility for communication to ever take place. Admittedly, in most its instances in nature, communication is achieved through the use of a common underlying code which allows the straightforward transmission of a message to an audience. Take for example the classic case of honeybees: A bee lets its beehive know the whereabouts of a good supply of nectar by moving its body in some particular manner. In this setting, communication occurs through the encoding of the information that is to be communicated (i.e., the location of the nectar) into a specific "dance", which the audience bees are able to decode through the implementation of an identical copy of the code that the communicating bee is using. From a first look, human verbal communication can also be treated with a similar rationale, according to which the existence of a common linguistic code between two interlocutors suffices on its own to guarantee their successful communication. In such an account, all that the speaker has to do is encode her message into a natural-language sentence and utter it. The hearer will in turn decode the utterance's meaning and faithfully reconstruct in this way the speaker's original message.

Though it may seem intuitively adequate, this scenario actually falls short of holistically explaining verbal communication. Largely due to Grice's (1989) pioneering work, it has been generally acknowledged that human communication is not entirely amenable to the aforementioned code-based treatment, but is essentially achieved through the expression and recognition of intentions. From this perspective, upon recognizing the speaker's intention to communicate a message to him and on top of decoding her communicated utterance, the hearer needs to inferentially construct the speaker-intended meaning on the basis of the recovered decoded content and the context of utterance. In the Gricean tradition, the paradigm output of this inferential process is an *implicature*, a proposition that is purposefully conveyed by the speaker's utterance without being part of its decoded meaning and which the hearer is expected to infer for the speaker's communicative intention to be fulfilled.

In its original exposition, Grice's argument did not straightforwardly compromise the validity of the aforementioned code-based account of verbal communication, but rather added an inferential layer to the decoding process that was customarily thought to take place during the comprehension of an utterance. This essentially brings us to the currently most common understanding of the semantics/pragmatics distinction and, according to which, semantics is the study of the linguistic code alone and, thus, responsible for providing us with the public meaning of sentences while pragmatics deals with the inferential processes that are needed for the construction of the speaker-intended implicated meaning when sentences are uttered in context. It is with respect to this view that one of the most heated debates in modern linguistics has arisen. For quite some time now, several formidable minds have been engaged with questions regarding the necessity and magnitude of contextual intrusions in the delineation of an utterance's explicitly expressed propositional content and the debate has been perpetuated through the argumentation of scholars belonging to two conflicting camps of thought, which Recanati (2004) neatly locates in the traditions of *Literalism* and *Contextualism* respectively. Roughly speaking, the literalist contends that natural language sentences carry truth-conditional content on their own while the contextualist argues that it is only in the context of its utterance that a sentence can be assigned a determinate truth value. Effectively then, the point of disagreement between the two doctrines amounts to whether fully propositional semantic content can exist in isolation from contextualization.

Naturally, in line with the general fascination of the contemporary linguist with formal approaches to semantics, it comes as no surprise that literalism is currently the dominant position with respect to the debate at hand. In their recent attack against contextualism, Cappelen and Lepore (2005) take the literalist position, which is clearly represented in their positive theory of Semantic Minimalism, to be only commonsense, given the classic Fregean claim that in communication we share thoughts. As they characteristically note, "if communicated contents are restricted to (or, essentially tied to) specific contexts of utterance, then it is hard to envision how speakers who find themselves in different contexts can communicate" (2005: 153). Provided then that contexts are personal and cannot be shared in the code-like manner that semantic contents are hypothesized to be shared across interlocutors, Cappelen and Lepore view the proposition expressed by a sentence irrespective of its context of utterance as our sole "minimal defense against confusion, misunderstanding, mistakes" during verbal communication (2005: 185). Therefore, Semantic Minimalism advocates that, in their communicative practices, both the speaker and the hearer mentally access not only identical, but also fully propositional semantic contents of the utterances utilized.

One of the frameworks that Cappelen and Lepore consider to be in direct opposition to their regime is Relevance Theory (Sperber and Wilson 1995). This is of course hardly unjustified, since relevance theorists hold a radical version of the linguistic underdeterminancy thesis, according to which, "linguistically encoded meaning *never* fully determines the intended proposition expressed" (Carston 2002: 49, emphasis in original). From the relevance-theoretic perspective, during comprehension, an utterance's semantic content might be indeed decoded in the aforementioned sense, but the outcome of this decoding process, the utterance's *logical form*, always needs considerable contextual input to gain full propositional status. In this respect, the inferential computation of the speaker's intentions while a logical form is recovered becomes crucial for the construction of an utterance's explicitly expressed content by the hearer and the product of this *parallel* and *simultaneous* operation of decoding and inference is a pragmatically enriched, yet properly truth-evaluable, proposition, which relevance theorists dub *explicature.*

Clearly, this approach is largely incompatible with Cappelen and Lepore's account. Its basic shortcoming, as the proponents of Semantic Minimalism themselves isolated it recently (Cappelen and Lepore 2007), is that it jeopardizes a central aspect of verbal communication, that is, the relation of identity that, according to Cappelen and Lepore, needs to exist between the basic proposition expressed by the speaker and the one constructed by the hearer for communication between them to take place. Indeed, by allowing context to intrude in the delineation of an utterance's explicit content, "we need recognize only speaker-relative content and listener-relative content and a relation of *similarity* holding between these two contents" (Bezuidenhout 1997: 198, emphasis in original). And this is a view that relevance theorists certainly share with most contextualists. For Cappelen and Lepore, however, this position is seriously flawed:

[…] there's no non-trivial sense of 'similarity' in which the explicatures arrived at by using the [relevance-theoretic comprehension procedure…] will be similar to the proposition that the speaker intended to communicate. They will, if R[elevance] T[heory] is correct, be developments of the same logical form LF, but an LF can be developed into *radically different* propositions. (Cappelen and Lepore 2007: 117)

Noticeably, this is hardly a new criticism. As Fodor and Lepore have also argued, "the kind of explanations that semantic theories are supposed to give would not survive substituting a similarity-based notion of content for an identity-based notion" (Fodor and Lepore 1999: 382), which is on its own particularly problematic since "nobody has the slightest idea how to construct the required notion of content similarity" (Fodor and Lepore 1999: 382).

Against this background, in the following section, I will defend the contextualist orientation by providing an overview of some basic sources of linguistic underdeterminancy, as these are pinpointed in the relevance-theoretic literature. Then, I will briefly present the basic tenets of Relevance Theory with a view to show that the comprehension procedure that it puts forth realistically addresses the on-line process by which interpretation takes place, safeguarding the validity of its outcomes in a non-trivial way. In the remainder of this paper, I will turn to the relevance-theoretic framework itself, evaluating further its conception of semantics. Discussing the implications that this conception carries with respect to lexical/conceptual content, I will finish my argumentation by proposing that relevance theorists need to recognize the context-sensitivity of semantic content even more cordially than they currently do.

2. The linguistic underdeterminacy thesis

The basic motivation of the contextualist stance in general and of the relevance-theoretic framework in particular is the need to provide a psychologically plausible account of the processes that are involved in the communication of utterances. Conversely, the roots of literalism are located in the early proposals of *ideal language philosophers*, like Frege, Russell, and Tarski, who sought to investigate meaning from the perspective of formal logic and "were not originally concerned with natural language, which they thought defective in various ways" (Recanati 2004: 1). For instance, Frege argued for "a sharp separation of the psychological from the logical, the subjective from the objective" (Frege 1997: 90), which indeed rendered natural language defective for his purposes, language being clearly a "mixture of the logical and the psychological" (Frege 1997: 243).

In this setting, formal approaches to semantics aim at examining necessary truths and essentially maintain that "being true is quite different from being held as true, whether by one, or by many, or by all, and is in no way to be reduced to it" (Frege 1997: 202). In contrast, contextualist accounts are specifically concerned with the "intuitive truth-conditions" of an utterance (Recanati 2004: 24), that is, with the conditions under which what an utterance asserts can be held to be true by its hearer. From this perspective, the context-independent truth-conditional content that formal semanticists assign to a sentence appears to be the result of a "principle which has absolutely no bearing on human psychology" (Carston 1988: 165). However, Cappelen and Lepore (2005: 176–189) attempt to rebut this criticism by arguing that their minimal propositions are psychologically real and, therefore, considered by the hearer during the interpretation process. Reinstating the contextualist criticism, in what follows, I will illustrate how the explicitly expressed propositional content of an utterance is customarily much more context-dependent than Cappelen and Lepore wish to persuade us.

To begin with, even Cappelen and Lepore acknowledge that a proper theory of semantics needs to incorporate at least some minimal context-sensitivity. It is by now commonly accepted that there exists a number of linguistic expressions, customarily called *indexicals*, which are clearly context-dependent. Even intuitively, expressions like "she," "here," "now" have a different truth-conditional contribution on most occasions of their use. Against this background, the basic tenet of Semantic Minimalism becomes that "the semantic content of a sentence S is the proposition that all utterances of S express (*when we adjust for or keep stable the semantic values of the obvious context sensitive expressions in S*)" (Cappelen and Lepore 2005: 2–3, emphasis my own). Based on Kaplan (1989), Cappelen and Lepore provide a list of these obviously context-dependent indexicals, assuming that these are merely exceptions to the rule of semantic content autonomy. In my view, this is hardly the case, given the frequency with which we employ such expressions – especially pronominals – in our everyday communicative practices, but, in all fairness, I will grant Cappelen and Lepore their window of doubt, since there is a sense in which the context-sensitivity of indexicals is special. Even in the contextualist literature, indexical expressions are clearly distinguished, since in their case it is the lexical item itself that mandatorily triggers the inferential process by which it will be assigned a determinate semantic value. And this process, often called *saturation*, is qualitatively different from the corresponding process of *modulation*, that is, the context-driven process by which the semantic content of an utterance is freely enriched in order to gain full propositional status.

Nevertheless, even if we accept Cappelen and Lepore's argument that the linguistically triggered context-sensitivity of a small set of expressions is indeed exceptional, the problems for Semantic Minimalism are not eradicated. In order to show this, I will now provide a–necessarily schematic–summary of some basic sources of linguistic underdeterminancy, as these are discussed at length within the relevance-theoretic framework by Carston (2002: 15–93).

In this respect, Semantic Minimalism may be prepared to accommodate the obvious context-dependence of indexicals, but there are still further cases of linguistic expressions whose reference cannot be determined without resorting to the context of utterance and which Cappelen and Lepore do not include in their list of context sensitive expressions. For one, much like indexicals, proper names also refer to different individuals in different contexts (Sperber and Wilson 1995). Similarly, as Higginbotham notes (1988), the assignment of reference to "incomplete" definite descriptions–as in (1)–and "specific indefinite" descriptions–as in (2)–can correspondingly be problematic for traditional truth-conditional accounts of semantics:

(1) The president (whoever s/he is) is in trouble.
(2) If a certain blonde calls, pass her through directly to me.

Accordingly, even definite descriptions with no indexical element whatsoever can be taken to be context-dependent. As Recanati discusses (1987, 1996), in every communicative act there is a *domain of discourse*, "with respect to which the speaker presents his or her utterance as true" (Recanati 1987: 62), and on which the reference of a definite description always depends. In order to illustrate this argument, let's consider the case of Ann's father uttering (3) to Ann during her graduation ceremony in 2008:

(3) Mum will be pleased to see that the current prime minister of Britain attended your graduation.

Imagine now that both Ann and her father know that Ann's mother believes that Blair is still the prime minister of Britain when he is not. In this case, Ann's father uses the definite description "the current prime minister of Britain" to refer to Blair, and Ann is respectively expected to interpret it with respect to her mother's belief system. In this particular setting then, the domain of discourse in which Ann will decode her father's utterance is not the actual world, as literalist accounts of semantics would predict. Along these lines, it becomes possible that the relativity of reference to domains of discourse can actually apply to all definite descriptions, even the ones that seem "the least likely to yield to the general context-dependence thesis" (Carston 2002: 38). Finally, the context sensitivity of an utterance's propositional content is also apparent in the case of lexical or syntactic ambiguity. For instance, in the tradition of semantics, the problem of lexical ambiguity is usually tackled with by positing what Pustejovsky (1995) calls a *sense enumerative lexicon*, which comprises different lexical entries for each sense that an ambiguous lexical item is considered to have. In this setting, the context of utterance again plays the essential role of disambiguating which of the two semantic entries of, say, the word *bank*, its every use points to.

Furthermore, the linguistic string employed in an utterance might still fall short of determining a full proposition "even after all necessary reference assignments and disambiguations have taken place" (Carston 2002: 22). Consider Carston's examples in (4) to (8) that follow:

(4) Paracetamol is better.   [than what?]
(5) It's the same.   [as what?]
(6) She's leaving.   [from where?]
(7) He is too young.   [for what?]
(8) It's raining.   [where?]

What is obvious here is that for the proposition expressed by these utterances to be assigned a determinate truth value, certain missing elements, often referred to as *unarticulated constituents*, need to be contextually supplied. And regardless of whether these constituents are treated as hidden indexicals (Stanley 2000) or as contextually derived elements of the proposition expressed (Recanati 2002), the context manages in both cases, albeit in different ways, to intrude in the truth-conditional content of an utterance.

Another source of linguistic underdeterminancy can be found in instances where a lexical item's scope is left unspecified from the semantics of the item itself. In this setting, Gross (2001) discusses the "part" context sensitivity of adjectival predicates, as this is exemplified in (9):

(9) The book is black.

It should be straightforward that in this case the meaning of "is black" is semantically underdetermined in relation to the property of the book that it refers to (the cover, the dominant part of the cover, the pages, etc). And admittedly, context-sensitivity of this type carries over to other expressions as well, such as verbs, nouns etc., as is evident in (10):

(10) John finished the book.

Again, the semantics of "finish" cannot on its own provide us with a full propositional content for this utterance, as John might have finished reading the book, writing it, binding it, and so on and so forth.

Finally, a related source of linguistic underdeterminacy can be located in instances where a lexical item's literal meaning, what semantic theory takes to be its encoded content, needs to be pragmatically adjusted for the communicator's intended meaning to be constructed by the hearer. Consider the cases in (11) and (12):

(11) Mary has a temperature.
(12) The fridge is empty.

It would be difficult to come up with the propositions explicitly expressed by these utterances in certain – and most likely familiar – contexts, if we did not contextually enrich the semantic content of *temperature* and *empty*. In (11), Mary's temperature can be easily attributed the narrower interpretation of "a high temperature" rather than its literal one, which would apply to Mary as it does to all living organisms. Similarly, in (12), the fridge will probably not be interpreted as being totally empty, but rather insufficiently filled with the goods that are needed by a household on a daily basis. In this case, the encoded content of "empty" needs to be broadened for its more precise intended interpretation to be yielded.

Although my overview here has been inevitably brief, I believe that it can suffice to show that the underspecification of semantic content does not evaporate once indexicality is recognized. In this respect, the literalist argument for the existence of context-independent basic propositions that are identically shared among interlocutors becomes particularly shaky, as contextual intrusions are, on multiple occasions, indispensable for the determination of an utterance's explicitly expressed proposition. Therefore, endorsing the linguistic underdeterminancy thesis, at least from the relevance-theoretic perspective, is more than a merely contingent matter: "underdeterminancy is an essential feature of the relation between linguistic expressions and the propositions (thoughts) they are used to express" (Carston 2002: 29).

3.   The relevance-theoretic comprehension procedure

As the previous section showed, contextual intrusions in the proposition explicitly expressed by an utterance are so frequent that they practically prevent the hearer from recovering a basic proposition that will be identical in content to the one that the speaker intended to communicate. On the face of this impossibility, it becomes obvious that the relation of similarity that contextualism holds to exist between the speaker-uttered and hearer-constructed contents essentially constitutes the only realistic alternative in a psychologically plausible discussion of our communicative practices. Indeed, for the relevance theorist, it is "neither paradoxical nor counterintuitive [...] that communication can be successful without resulting in an exact duplication of thoughts in communicator and audience" (Sperber and Wilson 1995: 193). However, as I noted above, Cappelen and Lepore reject this position, for it leaves open the possibility that the hearer will construct a radically different explicature from the one that the speaker might have intended to communicate. In this section, I will argue that even though the inferential process by which logical forms are developed into full propositions can in principle lead to this result, if we take the relevance-theoretic proposals seriously, it becomes highly unlikely that this will often happen in practice.

   To begin with, it seems necessary to note that Relevance Theory is essentially a theory of cognition and mental processing and that this is how it manages to shed light on how verbal communication realistically takes place. Using the term in a technical sense, Sperber and Wilson explore relevance as a psychological property of cognitive input to mental processing. In the current setting then, this input can be identified with a communicated utterance that needs to be processed in order to construct its intended meaning. From the relevance-theoretic perspective, the degree in which an utterance will be relevant to its hearer, and thus worth processing, depends on a balance of cognitive effects and processing effort:

Relevance of an input to an individual

a. Other things being equal, the greater the positive cognitive effects achieved by processing an input, the greater the relevance of the input to the individual at that time.

b. Other things being equal, the greater the processing effort expended, the lower the relevance of the input to the individual at that time.

While the notion of processing effort should be pretty self-explanatory in cognitive terms, the one of positive cognitive effects could do with some minimal elaboration. As should be evident by now, according to the relevance-theoretic framework, in order for the hearer to construct the message that the speaker intended to communicate to him, regardless of whether this message is explicitly expressed or implicated, he will need to utilize his general inferential abilities. In this respect, relevance theorists define non-demonstrative inference, as spontaneously used by us in our communicative practices, as a process that "starts from a set of premises and results in a set of conclusions which follow logically from, or are at least warranted by, the premises" (Sperber and Wilson 1995: 12–13). Effectively, this set of premises refers to the context, which comprises a number of assumptions that the hearer brings to the forefront of his attention, and against which he will infer the speaker-intended meaning during the comprehension of an utterance. The conclusions derived from the unification of the utterance with the context of interpretation then, will achieve positive cognitive effects if they bring about an improvement in the hearer's system of beliefs, by altering his already existing assumptions in a non-trivial way. Therefore, what follows from the above definition is that, the more an utterance improves the hearer's belief system and the less effort it requires in its processing, the more relevant it will be to him.

Having defined relevance in this way, Sperber and Wilson go on to discuss the notion's significance in relation to the way in which all mental computations take place and propose that considerations of relevance ultimately orchestrate the operation of our cognitive system in important respects. This is spelled out in their first or cognitive principle of relevance:

Cognitive Principle of Relevance

Human cognition tends to be geared to the maximization of relevance.

Following research in the domain of evolutionary psychology, Sperber and Wilson argue that, like most biological systems, human cognition should be viewed as a system that has evolved through processes of natural selection. Now, there is undoubtedly a vast number of stimuli surrounding us at any given moment, stimuli that can trigger our mental processing, but since we could not possibly attend to them all at the same time, we need to select the ones that we will focus on. In this respect, it is only in our interest to offer priority to those inputs from which we will benefit the most. In evolutionary terms, this creates an immediate pressure for our cognitive mechanism, a pressure to which it can be thought to have responded over time by adapting into a system that gives priority to those inputs that will provide it with the largest gains. And actually this certainly seems to be a valid observation if we consider the fact that certain stimuli, like the sound of an explosion, automatically impinge on our attention. Similarly, the need for an all the more efficient operation would correspondingly create a pressure for the mind to minimize the effort that it will have to spend in the processing of such particularly useful stimuli. Here, a case in point would be communicative stimuli, such as linguistic utterances, which we are normally able to comprehend, performing complex inferences, within only a few milliseconds. Against this background, it seems reasonable to assume that the human cognitive system has a natural tendency to attend to the most beneficial stimuli and allocate its resources in such a way that this processing will come as effortlessly as possible. And Sperber and Wilson's notion of *maximal relevance* can be directly implemented in the technical exposition of this argument, as it straightforwardly captures the idea that the search for the greatest possible effects for the least possible effort has been instrumental in our cognitive system's evolution.

Turning to communication now, Sperber and Wilson discuss communicative stimuli employing the same cognitive definition of relevance and positing in their argumentation a dedicated inferential processor, whose domain is purposefully communicated stimuli, and which automatically computes the full set of cognitive effects that a communicative stimulus produces in the mind of the addressee. Assuming then that human cognition is indeed geared towards maximal relevance, as the first principle of relevance predicts, and that communicative stimuli automatically pre-empt our attention, an interesting picture emerges. For one, it makes sense to suppose that an addressee can automatically expect a communicative stimulus to provide him with some adequate cognitive effect or else he would have the choice of not processing it. Then, given that the central goal of communication is to understand and be understood, it seems to be "in the communicator's [...] interest both to do her best and to appear to be doing her best to achieve this [...] goal" (Sperber and Wilson 1995: 268) by planning her utterance accordingly, otherwise the addressee's relevance-oriented cognitive system would again not necessarily pay attention to her communicative stimulus, as it automatically does. In the relevance-theoretic framework, these two observations combined create a fundamental constraint for the way in which an addressee will *always* treat a communicative stimulus, a constraint that is captured in Sperber and Wilson's second or communicative principle of relevance:

Communicative Principle of Relevance

Every act of ostensive communication communicates a presumption of its own optimal relevance.

In this principle, what the *presumption of optimal relevance* amounts to is merely what was established above, that is, that an ostensively communicated stimulus will always be treated by an addressee as "relevant enough to be worth his effort to process it" and as being "the most relevant one compatible with the communicator's abilities and preferences." Therefore, since once we recognize a particular stimulus as *ostensively*, that is, deliberately, communicated to us, it is impossible for us to not process it, we will always process it as optimally relevant.

In this setting, a fundamental difference between maximal relevance that applies to cognition and *optimal relevance* whose presumption guides the processing of communicative stimuli arises: the former signifies the best possible balance between effort and effect while the latter's goal is the retrieval of adequate cognitive effects for no unjustifiable processing effort expenditure. Naturally, during communication, the cognitive effects that a stimulus will produce in the mind of the addressee will be deemed adequate once his derived interpretation can be accepted as the one that the communicator intended him to arrive at. In this respect, the second principle of relevance predicts that, since hearers cannot but process communicative stimuli as optimally relevant, the comprehension procedure will end when these satisfactory cognitive effects are produced. And regarding the on-line process by which these cognitive effects will be sought by the addressee, Relevance Theory predicts that it will automatically take place following a specific route:

Relevance-theoretic comprehension procedure

a. Follow a path of least effort in computing cognitive effects: Test interpretive hypotheses in order of accessibility.

b. Stop when your expectations of relevance are satisfied (or abandoned).

Admittedly, this comprehension procedure follows directly from the two principles of relevance that I have outlined above. For one, given the necessity of contextual intrusions in the delineation of an utterance's basic proposition, as the linguistic underdeterminancy thesis clearly demonstrates, if human cognition were not geared towards maximal relevance, a hearer would have to arbitrarily test an infinity of interpretive hypotheses before reaching the most satisfactory one; a task that he clearly has neither the time nor the cognitive resources to perform for every single communicative act. Therefore, accepting that his cognitive system will have a natural tendency to chunk particular pieces of information together in a relevance-boosting manner, so that he can go on and test these hypotheses in the order in which they become accessible to him, is "not just a reasonable thrift, [...but] an *epistemically* sound strategy" (Sperber and Wilson 1996: 532). Accordingly, as the presumption of optimal relevance creates in the addressee the expectation that the communicator will have planned her utterance so that its meaning will be constructed by him without any unnecessary processing effort expenditure, it makes sense to assume that he will stop processing its meaning once he comes across an interpretation which he finds satisfactory with respect to the communicator's intentions along the path of least effort.

Against this background, it becomes evident that the relevance-theoretic comprehension procedure is not merely a convenient assumption regarding the interpretation of some communicative stimulus, but follows from a series of independently motivated arguments regarding the evolution and organization of the human cognitive system and the rationality that underlies human communication. In turn, these arguments safeguard that, on most occasions, the outcome of the interpretation process, be it some explicature(s) or implicature(s), will not substantially depart from the communicator's original intentions, although it might well do, as customarily happens in cases of miscommunication.

4. The public availability of lexical meaning

Up to this point, I have demonstrated how the relevance-theoretic approach can successfully react to the literalist tradition that takes semantic knowledge to be capable of providing an utterance's explicitly expressed propositional content on its own. Having addressed the procedure by which it foresees that the inferential developments of an utterance's logical form will lead the hearer to the speaker-intended meaning in a psychologically realistic manner, I will now turn to assess the way in which the relevance-theoretic framework identifies lexical semantic content in its premises.

In this respect, Sperber and Wilson seem to hold a rather traditional view of semantics, according to which, an utterance's decoded content, its logical form, can be identified with its underlying sentence's semantic representation:

> By definition, the semantic representation of a sentence, as assigned to it by a generative grammar, can take no account of such non-linguistic properties as, for example, the time and place of utterance, the identity of the speaker, the speaker's intentions, and so on. The semantic representation of a sentence deals with a sort of common core of meaning shared by every utterance of it. (Sperber and Wilson 1995: 9)

From this statement, it becomes clear that the relevance-theoretic notion of logical form is essentially context-insensitive, which further suggests that Relevance Theory and Semantic Minimalism are not as incompatible as they might seem to be at first sight. As Wedgwood (2007) first noted and Carston (in press) asserts, both traditions pinpoint some semantic content upon which pragmatic inference effectively operates. Naturally, the relevance-theoretic conception of this encoded content departs substantially from the corresponding minimalist one, since its contextual development which occurs alongside its decoding, is indispensable for it to gain full propositional status.

However, no matter how schematic or propositionally incomplete a sentence's logical form might be considered from this perspective, it still comprises some content, the semantic content of the lexical items that constitute it, that cannot but be virtually context-independent as well. And following Sperber and Wilson's rationale, this encoded lexical meaning should be respectively viewed as the common core of meaning shared by every usage of the lexical item in abstraction from individual contexts of use. Therefore, even though relevance theorists argue that the hearer "entertains thoughts [and not...] semantic representations of sentences"

(Sperber and Wilson 1995: 193) at the propositional level, they still maintain that at the lexical one he will need to entertain an expression's semantically encoded content before deciding whether to contextually enrich it or not. In the setting of psychological plausibility in which Relevance Theory purports to investigate linguistic communication then, a fundamental question immediately presents itself: How are we to realistically pinpoint what this context-independent meaning is with respect to some linguistic expression?

To begin with, it seems necessary to briefly illustrate the Fodorian background on which the relevance-theoretic conception of semantics squarely lies. In this respect, much like Fodor, relevance theorists distinguish between "linguistic" and "real" semantics in the following way:

> linguistic semantics [...] could be described in statements of the form *'abc' means (= encodes) 'ijk'*, where 'abc' is a public-language form and 'ijk' is a Mentalese form (most likely an incomplete, schematic Mentalese form) [while...] 'real' semantics [...] explicates the relation between our mental representations and that which they represent [...] and whose statements may take the form *'hijk' means (= is true iff) such-and-such*. (Carston 2002:58)

In this sense, Relevance Theory adopts the further Fodorian argument that language "inherits its semantics from the contents of beliefs, desires, intentions, and so forth that it's used to express" (Fodor 1998: 9). By definition then, the semantic content of a lexical item is inherited by the "real" semantics of its associated mental concept. And regarding this "real" semantics, Fodor adopts a purely externalist perspective, according to which, the true uses of a concept, such as CAT, are caused by actual cats.

Similarly, when addressing lexical meaning, Sperber and Wilson begin their discussion by positing mental concepts as the direct cognitive counterparts of lexical items. In this respect, "the 'meaning' of a word is provided by [its...] associated concept (or, in the case of an ambiguous word, concepts)" (Sperber and Wilson 1995: 90), where a concept can be viewed as a stable address in memory comprising three entries. Of these entries, the most straightforward to approach seems to be the *lexical* one, which is taken to encompass syntactic and phonological information about the lexical item that encodes a concept in natural language. Then, a concept's *logical* entry consists of a set of inferential rules, commonly known as *meaning postulates*, whose function is to provide the inferential processor with sets of premises and conclusions that "capture certain analytic implications of the concept" at hand (Carston 2002: 321). Finally, the *encyclopedic* entry of a concept contains general information that we individually hold with respect to its denotation; information that can be arbitrarily stored in the form of full propositions, assumption schemas or mental images.

From the relevance-theoretic perspective, the existence of both a lexical and a logical entry in the same conceptual address is deemed necessary given each concept's double contribution to decoding and inference during utterance interpretation. As Sperber and Wilson themselves observe, "recovery of the content of an utterance involves the ability to identify the individual words it contains, to recover the associated concepts, and to apply the [inferential...] rules attached to their logical entries" (Sperber and Wilson 1995: 90). In this respect, it is the lexical entry of a word that activates the meaning postulates that are included in the logical entry of its encoded concept, which in turn enter the inferential processor one-by-one during an utterance's decoding into its logical form. Regarding the distinction of the logical from the encyclopedic entry now, Relevance Theory argues that it essentially corresponds to the traditional analytic/synthetic distinction. Along these lines, a concept's logical entry is perceived to comprise inferential rules that express analytic, and thus necessary, truths with respect to the concept's meaning while its encyclopedic entry contains chunks of memorized information that can optionally enter the inferential processor contributing to contextual enrichments of the concept's encoded content.

In this respect, it seems to me that the most salient way in which we can construe a lexical item's encoded content from the relevance-theoretic perspective is by identifying it with the inference rules that are included in its associated concept's logical entry. And the following argument, put forth by Sperber and Wilson themselves, certainly points to that very direction:

> Encyclopaedic entries typically vary across speakers and times: we do not all have the same assumptions about the Napoleon or about cats. They are open-ended: new information is being added to them all the time. There is no point at which an encyclopaedic entry can be said to be complete, and no essential minimum without which one would not say that its associated concept had been mastered at all. Logical entries, by contrast, are small, finite, and relatively constant across speakers and times. There is a point at which a logical entry for a concept is complete, and before which one would not say that the concept had been mastered at all. (Sperber and Wilson 1995: 88)

Therefore, by maintaining that a lexical item's semantic content is context-insensitive and thus stable across individuals and times, relevance theorists cannot but equate it with its associated concept's set of meaning postulates, since the information included in its encyclopedic entry is highly individualistic and essentially private. However, given the aim of psychological plausibility with which Relevance Theory was developed to begin with, this seems like a rather counterintuitive move.

Following Quine (1951), it is now commonly accepted that purely logical analytic truths do not exist and that our intuitions of analyticity are empirically driven. In this respect, a statement might indeed appear to be *a priori* true, but is always amenable to revision, once further beliefs to which it closely relates are correspondingly modified. In light of this argument, even Fodor himself, who originally introduced meaning postulates in the discussion of conceptual content, was led to question their theoretical usefulness, eventually abandoning them from his theoretical investigation of concepts (Fodor 1998). Therefore, it seems that the conceptual content that relevance theorists identify as semantic crucially rests on an argument regarding analyticity that was ill-founded and psychologically unrealistic to begin with. Even so, Relevance Theory cannot discard meaning postulates as straightforwardly as Fodor did, since a central aspect of their account of utterance interpretation crucially rests on their very application. Without logical inferential rules, an utterance's decoding into its logical form would be practically unfeasible.

Against this background, the only way in which the relevance-theoretic framework could preserve its current account without compromising its fundamental aim of psychological plausibility is by substituting its traditional conception of meaning postulates with a more realistic counterpart. Such an attempt has been recently made by Horsey (2006), who proposes a replacement of the problematic notion of analyticity with that of *psychosemantic* analyticity, according to which, "while the majority of our psychologically represented inference rules are no doubt veridical, this is by no means necessary" (Horsey 2006: 74). And given Quine's argumentation, it certainly becomes clear that the relevance-theoretic framework needs to incorporate psychosemantic analyticity so construed in its premises.

However, by necessarily psychologizing analyticity, Relevance Theory cannot but respectively psychologize encoded conceptual contents, abandoning its view of lexical semantics as effectively context-independent. In the original relevance-theoretic account, it was only by means of the externalist scenario with respect to the way in which we acquire content-constitutive inferential rules that the identity of semantically encoded contents across individuals and times could be guaranteed. In this setting, if the logical entry of the concept CAT would include the inference rule Φ CAT $\psi \rightarrow \Phi$ ANIMAL ψ, it would follow that the context-independent semantic content of the word 'cat' would nomologically contain at least the information that cats are animals in all its mental instantiations across individuals. By accepting psychosemantic analyticity, however, there can be plenty of cases where two individuals' meaning postulates with respect to the same concept will vary. To use an example that Horsey (2006: 74) himself discusses, in the new picture, we can accept the meaning postulate Φ WHALE $\psi \rightarrow \Phi$ FISH ψ as content-constitutive of Mary's concept WHALE, even though it is not veridical in relation to the real world. However, there will certainly be other individuals whose logical entry of the same concept will not contain the very same meaning postulate, substituting it for Φ WHALE $\psi \rightarrow \Phi$ MAMMAL ψ. In this respect, it becomes evident that, if a concept's meaning postulates do not express necessary truths about the concept's extension in the actual world, but rather constitute psychological constructs that capture what an individual perceives to be necessary truths regarding it, then by definition the logical entry of this concept cannot be publicly shared. Therefore, since lexical semantic contents can typically vary across individuals, the very notion of some "common core of meaning shared by every usage of a lexical item in abstraction from individual contexts of use" that Relevance Theory endorses becomes practically vacuous. Naturally, this does not necessarily entail that there is no lexical semantic content as such, but rather that, much like its propositional counterpart, this content also succumbs to the radical linguistic underdeterminacy thesis in the communicative setting.

In order to work out the main implication that this conclusion carries for the relevance-theoretic account, I have to first briefly illustrate the on-line process by which Relevance Theory addresses contextual enrichments at the lexical level. Following Barsalou (1983, 1987) and his experimental evidence regarding the flexibility with which we can entertain temporary mental constructs that arise in particular contexts, relevance theorists suggest that during utterance interpretation an encoded concept might get pragmatically adjusted forcing us to construct an *ad hoc* concept in its place. In this respect, after the decoding process provides the inferential processor with the meaning postulates of the concept associated with a word, the inferential processor might either accept this encoded meaning as it stands or contextually adjust it, following the relevance-based comprehension procedure, in an attempt to provide certain positive cognitive effects that will satisfy the hearer's expectation of optimal relevance.

In this sense, from the relevance-theoretic perspective, all the lexical meanings that the hearer will construct during the interpretation of an utterance can be viewed as "outcomes of a single pragmatic process which fine-tunes the interpretation of virtually every word" (Wilson 2004: 344). Given then the current proposal that the semantic content of a linguistic expression is best viewed as individualistic and cannot be publicly shared, the tantalizing possibility that all communicated lexical meanings essentially correspond to *ad hoc* concepts seems to present itself. And that is because the hearer cannot but contextually enrich a lexical item's semantic content against the speaker-intended context, as if he does not do so, there is always a chance that his individualistic semantic content for that lexical item will not be similar enough to the one that the speaker intended to communicate.   Even though this is a position that like most relevance theorists I am currently reluctant to explicitly defend, since it would require a complete reidentification of the theoretical notion of logical form, it certainly seems to provide fruitful ground for further investigation (for a more detailed argument along these lines, see Assimakopoulos 2008). For the time being, I believe that my argumentation should suffice for my current purposes of showing that lexical semantic content is much more context-sensitive than Relevance Theory currently maintains. Against this background, it seems that relevance theorists should eventually open up to the possibility of incorporating a more radical version of contextualist semantics in their framework's basic exposition.

5.   Concluding remarks

One of the basic assumptions of most accounts of communication is that for communication to be successful some sharedness of information between the speaker and the hearer is necessary. Likewise, from the relevance-theoretic perspective, "communication requires some degree of co-ordination between communicator and audience on the choice of a code and a context" (Sperber and Wilson 1995: 43). A basic innovation of the relevance-theoretic approach with respect to this issue has been that it does not take this coordination to be a given as traditional code theories of communication do. Rather, it takes the most plausible route that interlocutors inferentially establish common ground during their communicative practices. As I have shown in this paper, inference does not realistically kick in after the identification of some minimal semantic content that is publicly available to each and every one of us, but essentially operates *in parallel* to this decoding process, in ways that might be even more radical than relevance theorists currently acknowledge. Indeed, by endorsing the literalist perspective and its corresponding idea of semantic content identity between interlocutors, we could in principle explain communication rather easily. However, in doing so we would essentially turn our back on psychological plausibility; prioritizing psychological explanation over theoretical convenience, it becomes crucial that we start to seriously entertain the possibility that discussions regarding the inferential attribution of intentions might be much more instrumental for the study of linguistic semantics than customarily perceived.

References

Assimakopoulos, Stavros
    2008   Logical Structure and Relevance. Ph.D. diss., University of
             Edinburgh.
Barsalou, Lawrence W.
    1983   Ad hoc categories. *Memory and Cognition* 11: 211–227.
    1987   The instability of graded structure. Implications for the nature of
             concepts. In *Concepts and Conceptual Development: Ecological and
             Intellectual Factors in Categorization*, Ulric Neisser (ed.), 101–140.
             Cambridge: Cambridge University Press.
Bezuidenhout, Anne
    1997   The communication of *de re* thoughts. *Noûs* 31:197–225.
Cappelen, Herman, and Ernie Lepore
    2005   *Insensitive Semantics. A Defense of Semantic Minimalism and
             Speech Act Pluralism*. Malden: Blackwell.
    2007   Relevance theory and shared content. In *Pragmatics*, Noel Burton-
             Roberts (ed.), 115–135. Basingstoke: Palgrave Macmillan.
Carston, Robyn
    1988   Implicature, explicature, and truth-theoretic semantics. In *Mental
Representations. The Interface between Language and Reality*, Ruth M. Kempson
             (ed.), 155–181. Cambridge: Cambridge University Press.

2002   *Thoughts and Utterances. The Pragmatics of Explicit Communication*. Oxford: Blackwell.

In press   Linguistic communication and the semantics/pragmatics distinction. To appear in *Synthese*.

Fodor, Jerry A.
1998   *Concepts. Where Cognitive Science Went Wrong*. New York: Oxford University Press.

Fodor, Jerry A., and Ernie Lepore
1999   All at sea in semantic space. Churchland on meaning similarity. *The Journal of Philosophy* 96: 381–403.

Frege, Gottlob
1997   *The Frege Reader.* Michael Beany (ed.), Oxford: Blackwell

Grice, H. Paul
1989   *Studies in the Way of Words*. Cambridge, MA: Harvard University Press.

Gross, Steven
2001   *Essays on Linguistic Context-sensitivity and its Philosophical Significance*. New York: Routledge.

Higginbotham, James
1988   Contexts, models, and meanings. A note on the data of semantics. In *Mental Representations. The Interface between Language and Reality*. Ruth M. Kempson (ed.), 29-48. Cambridge: Cambridge University Press.

Horsey, Richard
2006   The Content and Acquisition of Lexical Concepts. Ph.D. diss., University College London.

Kaplan, David
1989   Demonstratives. In *Themes from Kaplan*, Joseph Almog, John Perry, and Howard Wettstein (eds.), 481–563. New York: Oxford University Press.

Pustejovsky, James
1995   *The Generative Lexicon*. Cambridge, MA: MIT Press.

Quine, Willard V.
1951   Two dogmas of empiricism. *The Philosophical Review* 60: 20–43.

Recanati, François
1987   Contextual dependence and definite descriptions. *Proceedings of the Aristotelian Society* 87: 57–73.

1996   Domains of discourse. *Linguistics and Philosophy* 19: 445–475.

Recanati, François. 2002. Unarticulated constituents. *Linguistics and Philosophy* 25: 299–345.

2004   *Literal Meaning*. Cambridge: Cambridge University Press.

Sperber, Dan, and Deirdre Wilson
   1995   *Relevance. Communication and Cognition.* 2d ed. Oxford:
            Blackwell.
   1996   Fodor's frame problem and relevance theory. *Behavioral and Brain*
            Sciences 19: 530–532.
Stanley, Jason
   2000   Context and logical form. *Linguistics and Philosophy* 23: 391–434.
Wedgwood, Daniel
   2007   Shared assumptions. Semantic minimalism and relevance theory.
            Journal of Linguistics 43: 647–681.
Wilson, Deirdre
   2004   Relevance and lexical pragmatics. *UCL Working Papers in*
            Linguistics 16: 343–360.

The construction of epistemic space via causal connectives

Montserrat González and Montserrat Ribas

1. Introduction

Causal connectives are typically described as elements that help establish a logico-semantic relationship of cause and effect between first and second segment in a sentence. Thus, the connective *because* in *I took the umbrella this morning because it was raining* contributes to create a semantic type of discourse coherence relation between the two segments due to their propositional content, i.e. de locutionary meaning of the two segments (Sanders 1997: 122); the connection established by the connective is part of our world knowledge and of propositional logic. On the other hand, the connective *because* in *He's ill, because he doesn't look well* contributes to create a pragmatic type of coherence relation between the two segments due to the illocutionary meaning of one or both of the segments, where "the state of affairs in the second segment is not the cause of the state of affairs in the first segment, but the *justification* for making that utterance." (Sanders 1997: 122). Based on Sanders (1997) notion of *source of coherence* and on his *Basic Operation Paraphrase Test* applied to causal relations, this article aims to show that most causal relations set up by *because* in spontaneous oral discourse are pragmatic. In addition, the study relates such use of the connective with evidential marking (Chafe 1986; Ifantidou 2001). Presenting data from a Catalan oral corpus of interviews to women that occupy outstanding positions in the political, economic, and cultural Catalan society, we hypothesize that when the speaker makes use of pragmatic *because* (Spanish *porque*; Catalan *perquè*) the source and mode of knowledge from which the cause-effect relationship is established becomes of primary importance, since causal relations originate from different sorts of evidence that the speaker interprets. The construction of coherence in an oral discourse presents striking differences from that found in a written piece of discourse. Speakers involved in a conversational exchange assign specific meanings to their utterances as the flow of the conversation develops, all determined by specific time, space, and participant type constraints that written discourses do not generally have. Furthermore, the type of genre (in this case, interviews) and the sort of questions posed to elicit the discourse under analysis play a key role. Findings suggest that causal structures are fundamental in the construction of knowledge or epistemic contextual spaces and that these have a direct influence on the way the listener processes the information. Finally, it is maintained that this has a direct effect on the creation of certain generalized social beliefs and on people's attitudes that are often unquestioned.

2. How linguistic cues aid in interpretation

In the last two decades, there has been a growing interest on linguistic cues (words, phrases) that help the speaker or writer signal intended intentions, thoughts and actions, facilitating interlocutors' correct interpretation of a text or discourse. By means of these cues, the listener or reader understands not only the propositional content of the message but also all that part which is related to procedural meaning, that is to say, all that has to do with the interpretive processing of the information (i.e., presuppositions, contextual assumptions and implicatures). There is a multiple array of terms behind the conceptualization of these linguistic units; the literature shows that one term can embrace a diversity of definitions and various terms can be used to refer to the same underlying notion; however, the most frequently used terms are *discourse markers* and *connectors* or *connectives*[1]. For the purpose of this study, we will use the term *connective*, commonly found in classifications related to argumentative discourse operations (causal, contrastive, temporal, reformulative, additive), and will focus on a specific type, *causal connectives*, more specifically on the Catalan causal connective *perquè*.

The notion of *visée argumentative,* adopted by Argumentation Theory (Ducrot 1983) from Bakhtine's (1977) work on poliphony, is crucial to understand argumentative connectives. This notion implies that certain linguistic elements of an utterance that carry propositional meaning are liable to lead towards specific conclusions that can be materialized differently, by means of locutionary or illocutionary acts (Ducrot 1983: 7). In this line of thought, the role of an argumentative connective is to carry out a specific argumentative relation (of cause and effect, reformulation, contrast, addition, temporal) from either a logico-semantic or a pragmatic perspective. The pragmatic role of connectives has been highlighted by numerous authors working on argumentative discourse relations that take into account not only propositional content but also speech acts and illocutionary meaning. See the Spanish examples provided by Briz (2001: 171) when discussing the two relation types. He illustrates them by means of causal *porque*.

(1) Ha ido al médico *porque* está enfermo
    EFFECT   CAUSE
    He's gone to the doctor because *he's ill*

(2) Está enfermo, *porque* ha ido al médico
    CLAIM   JUSTIFICATION
    He's ill, because *he's gone to the doctor*

As Briz points out, *porque* in (1) has a grammatical and semantic value; it works as a conjunctive device and connects two propositions related by an argumentative operation of cause and effect (pRq) whose linguistic framework is the sentence. On the other hand, *porque* found in (2) has a pragmatic value; the relationship it establishes between p and q is illocutionary since it introduces an argument that justifies the speaker's claim (He's ill); contrary to (1), its linguistic context in this case is the utterance. Briz refers to connectives working pragmatically as *pragmatic connectors* and to those which carry out a logico-semantic role as *syntacticosemantic connectors*. The latter allow a series of grammatical transformations (word order, negative and interrogative mode) that prove their different nature from the former (2001: 171–172).

The grammatical logico-semantic or the functional illocutionary value of the two sorts of argumentative connectives requires taking into account the textual properties of *cohesion* and *coherence*. Although there is not full agreement on the use of these two terms to identify two different underlying concepts, the latest contributions on the matter seem to conclude that *cohesion* and *coherence* are two distinct interrelated concepts. *Cohesion* relates to the surface sequential organization of a text, supported by the syntactic structure through such mechanisms as paraphrasing, repetition, parallel structures, pro-forms, ellipsis and inter- and intrasentential junction (Halliday and Hasan 1976). On the other hand, *coherence* is a cognitive process that implies configuration of concepts and knowledge of the world, both activated when processing a text (Van Dijk 1977, Beaugrande and Dressler 1981, Blakemore 1989). This cognitive view around the notion of coherence is shared by many linguists who work in the fields of pragmasemantics and artificial intelligence, and relate coherence to an interpretive process. According to Blakemore: "Even when two sentences are related by a cohesive tie, hearers have to go beyond their linguistic resources in order to recover an interpretation" (1989: 232). The issue of interpretability of a text or discourse is highly related to illocution and speech acts that accompany its propositional value. Thus, most of the times the understanding of a text, understood as a communicative piece of spoken or written information, requires going underneath its linguistic surface and looking for possible contextual effects, presuppositions and implicatures. This means that whereas *cohesion* is explicited by means of linguistic mechanisms and propositional development, *coherence* is inferred by the hearer or reader of a text. Next question that arises, then, is the following: how do these two textual properties relate to argumentative connectives? Taking the above discussion on connectives into account, we can easily establish a correlation between pragmatic argumentative connectives and the interpretive process underlying coherence, and logico-semantic argumentative connectives and the surface-level mechanisms of cohesion.

In the following section, we will see in more detail the categorization of coherence in terms of *source of coherence* (Sanders et al. 1993; Sanders 1997) related to causal connectives.

3.   Categorization of *source of coherence* and causal connectives

According to Sanders (1997: 119), people have or make cognitive representations of a discourse to account for text coherence and connectedness. Thus, a relation of cause and effect between two segments, for instance, can be inferred without any sort of linguistic cue and yet be understood as connected discourse (3), or it can be overtly explicit by means of a causal connective (4). Sanders concludes that this is possible because we establish different relations in discourse, one at the illocutionary level which concerns the speech act status of the segments, and another that considers their locutionary meaning, taking their propositional content into account.

(3)   My sister is not at home. Her car is not parked outside.

(4)   I switched on the light *because* it was dark.

Thus, the reason why (3) is coherent is because listeners understand the second segment as evidence for the claim in the first segment and not because there is a cause-effect relationship between the two as events taking place in the real world that form part of our world knowledge, as in (4): we know that when it is dark, we switch on lights (to see better), but the fact that the car is not parked outside does not *necessarily* mean that somebody is not at home. Note that the pragmatic relationship set up in (3) can also be established by means of a causal connective that makes explicit the claim-evidence illocutionary status of the utterance: "My sister is not at home *because* her car is not parked outside," and the same applies in (4) if we take out conjunctive *because*: "I switched on the light. It was dark." The presence or absence of the causal connective facilitates the interpretation and makes the relation (pragmatic or semantic) linguistically explicit, but it is not "compulsory." As previously mentioned, the coherence of these two sentences is found in the cognitive representation that we, as users of a language, have or make of them.

The way language users produce and understand discourse, viewed as cognitive representations, is the key research question of many scholars working on language processing. Sanders, Spooren and Noordman (1993) propose a taxonomy which aims to account for coherence discourse relations, after considering that most classifications were mere lists that could be extended endlessly (Martin 1992, Mann and Thompson 1988, inter alia). They describe them by means of four basic primitives or notions, one of which is *Source of coherence,* which explains if the relation between two segments is pragmatic or semantic: "A relation is semantic if the discourse segments are related because of their propositional content, i.e., the locutionary meaning of the segments." Thus, Sanders' example "Theo was exhausted *because* he had to run to the university" is a coherent sequence because the fact that running causes fatigue is part of our "world knowledge" (Sanders 1997: 122) "A relation is pragmatic if the discourse segments are related because of the illocutionary meaning of one or both of the segments. In pragmatic relations the CR [coherence relation] concerns the speech act status of the segments." And "Theo was exhausted, *because* he was gasping for breath" is a coherent sequence because the cause-consequence relationship determined by the connector is based on a "real world link": "the state of affairs in the second segment is not the cause of the state of affairs in the first segment, but the justification for making that utterance." (Sanders 1997: 122).

Distinctions similar to Sander's "source of coherence" are *pragmatic* vs. *semantic* connectives (Van Dijk 1977; Briz 1994), *internal* vs. *external* uses of conjunctions and relations (Halliday and Hasan 1976; Martin 1992), *presentational* vs. *subject-matter* relations (Mann and Thompson 1988), *rhetorical* vs. *ideational* discourse markers (Redeker 1990), or *content* vs. *epistemic* and *speech act domain of language use* (Sweetser 1990). In any case, the semantic versus pragmatic dichotomy is a fundamental piece for understanding the nature of connectives[2]. Sanders et al. (1993: 94) claim that coherence "is not a property of the discourse itself but of the representation people have or make of it." Thus, what is coherent is not the discourse *per se* but the discourse *representation*. Sanders' *Basic Operation Paraphrase Test* applies the source of coherence concept to causal connectives (1997: 126), subject-matter of the present study. The test consists of two pairs of basic causal operations between two propositions that, by means of paraphrase formulations, show whether the relation is pragmatic or semantic. The first pair responds to a pragmatic relation; the second to a semantic one.

(i) a. the fact that P causes S.'s *claim/advice/conclusion* that Q;
(i) b. the fact that Q causes S.'s *claim/advice/conclusion* that P
xxx
(ii) a. the fact that P causes *the fact* that Q;
(ii) b. the fact that Q causes *the fact* that P

Thus, according to Sanders (1997: 126–127), in "I'm busy. You can take your own beer out of the fridge" the relation is pragmatic because "one of the paraphrases (i) corresponds best to the coherent relation as it is originally expressed in the text":

(i) The fact that I am busy *causes my advice* to take your own beer out of the fridge.
(ii) ?The fact that I am busy *causes the fact* that you can take your own beer out of the fridge.

In contrast, in "Theo was exhausted because he had to run to the university", the relation is semantic because "one of the paraphrases (ii) corresponds best to the coherent relation expressed in the text":

(i) ?The fact that Theo had been running *causes my claim* that he was exhausted.

(ii) The fact that Theo had been running *causes the fact* that he was exhausted.

According to Sanders, the two levels–semantic/locutionary/propositional and pragmatic/illocutionary/epistemic/speech act–are not strictly separable and not always clear-cut; pragmatic relations can be, the same as semantic ones, based on connections in the real world, but the distinction is that the relevant level of pragmatic connection is always illocutionary, even if it is linked to the locutionary one (1997: 123).

Finally, in relation to the way causal constructions operate and are grammatically constructed, Viana and Suïls (2002) introduce an interesting point to bear in mind, which is the Latin notion of *re* and *dicto* that differentiates the locutionary from the illocutionary ones (in coherence terms discussed above, semantic and pragmatic). The *re* causal constructions are those ruled by propositional logic; the subordinate clause introduces the cause of the locutionary content of what is sustained in the main clause, which presents the consequence or effect (5). On the other hand, the subordinate clause in the constructions of *dicto* describes the cause of the situation that is presented in the illocutionary content of the main clause[3] (6). See the examples presented by the authors (2002: 2941).

(5) Dinaven perquè tenien el plat a taula
   They had lunch because their meal was ready

(6) Dinaven, perquè tenien el plat a taula
   They had lunch, because their meal was ready

A third example that Viana and Suïls introduce illustrates a construction that may be confused by causal, which is the explicative (7).

(7) Dinaven, ja que tenien el plat a taula
   They had lunch, since their meal was ready

In the following section, we will present the relationship between Catalan causal connective *perquè* and epistemic space, understood from the point of view of evidentiality.

4. Causal connectives and evidential epistemic space

In general terms, there are four main basic relations between two discourse segments joined by an argumentative connective: additive, adversative, causal and temporal, all of them with their subcategories (Halliday and Hasan 1976, Martin 1992, Sweetser 1990, Fuentes 1996, Bateman and Rondhuis 1997, Briz 2001). They all make use of conjunction to express either *external* or *internal* relations:

> When we use conjunction as a means of creating text, we may exploit either the relations that are inherent in the phenomena that language is used to talk about [i.e. *external*], or those that are inherent in the communication process, in the forms of interaction between speaker and hearer [i.e. *internal*]; and these two possibilities are the same whatever the type of conjunctive relation, whether additive, adversative, temporal or causal. In fact we usually exploit both kinds. The line between the two is by no means always clear-cut; but it is there, and forms an essential part of the total picture. (Halliday and Hasan 1976: 241)

Both *external* and *internal* relations reflect the experiential and the interpersonal functions of language, the first understood as meanings that represent contents of the real world described in the text, and the second as meanings that represent the speaker's rhetorical intentions and interactions that take place in a communicative situation. However, under the heading of causal relations we may find subcategories such as *result, reason* and *purpose* that show blurred limits and are not always distinct and clear-cut. The internal and the external categories do not seem to apply then because causal relations tend to include a subjective reasoning or argument that often involves the speaker's evaluative perspective concerning his or somebody else's grounds for belief or motivation for expressing the belief (Halliday and Hasan 1976: 240; Martin 1992: 179). Sanders (1997: 124) discusses the three-domain model proposed by Sweetser (1990) that takes possible ambiguities on relations like the aforementioned into account. Her model proposes three possible interpretations: *content, epistemic* and *speech act*; the first, logico-semantic (external); the last two pragmatic (internal). See the *because* examples (8–10) with the three possible readings/interpretations proposed by Sweetser:

(8) John came back *because* he loved her.

(9) John loved her, *because* he came back.

(10) What are you doing tonight, *because* there's a good movie on?

In order to illustrate her three-domain theory, Sweetser paraphrases her examples. In (8) we have a content-domain: the real-world cause of John's coming back was his love for her; the conjunction is used to establish the relationship between the one event that causes the other in reality terms. In (9) we find a personal interpretation in the epistemic domain that can be paraphrased as: "My knowledge of John causes my conclusion/enables me to claim that he loved her". Finally, (10) could be paraphrased in speech act terms: "I'm asking you what you're doing tonight because I have a suggestion -to see a good movie" (Sanders 1997: 124).

The epistemic domain proposed by Sweetser is found at the pragmatic level of interpretation. The connection or relation holds, according to her, in the writer's or speaker's head; an argument is presented as a truthful claim and the person who makes it is involved in the communicative process. The *evidence* that the speaker has for making the claim is subjective (i.e., internal) and is based on observation or knowledge, as the paraphrasing of (9) above has shown. This point results of special interest since it allows us to establish a link between pragmatic epistemic domain of a causal relation and the notion of *evidentiality*, this understood as source of knowledge or evidence on which statements are based (Chafe 1986; Mithun 1986; Willett 1988; Wierzbicka 1994; Plungian 2001; Ifantidou 2001; Marín-Arrese 2004).

Evidential marking involves the speaker's or writer's stance in relation to source of information. It is often treated as a type of epistemic modality because the speaker estimates the chances that an utterance has of being true and expresses his/her degree of commitment to the truth of its propositional content. In contrast with deontic modality, the epistemic one focuses on the beliefs and knowledge of the person uttering the words, taking his/her own stance as point of reference. De Haan (1999:85) argues that whereas epistemic modality *evaluates* evidence, evidentials *assert* that there is evidence, without the speaker undergoing any kind of interpretive process. This evidence can be of different sorts: perceptual, inferential or by hearsay are the three broad categories. In relation to causal connectives, the inferential function is especially significant. Chafe (1986:265) suggests a relationship between different ways to acquire knowledge (*modes of knowing*) and the source or origin of such knowledge (*source of knowledge*), defining evidentials as devices that signal epistemology by coding the speaker's attitude toward his/her knowledge of the situation. See in Table (1) the classification and correlation that he proposes.

Table 1. Chafe's classification on evidentials

Modes of knowledge	Source of knowledge
BELIEF	[not clear]
INDUCTION / INFERENCE	Evidence
PERCEPTION	Senses
HEARSAY	Language
DEDUCTION	Hypothesis

Chafe's study on evidentiality (1986) was done in a corpus of conversational English. His data offer interesting qualitative information that highlights the specificities of spoken language in contrast with the written register. He points at the fact that "there are certain kinds of epistemological considerations that a writer has time to deal with, and a speaker typically does not"; speaking takes place "on the fly" and it is marked in terms of evidentiality with different degrees of reliability (1986:262–263). Figure (1) exemplifies and presents the two sorts of coherence relations, relating the pragmatic one to the different types of *modes of knowing* suggested by Chafe (1986). Next to Catalan *perquè*, we include also the Spanish and English conjunctions.

The two examples presented in Figure (1) show the difference, in coherence terms, between a *because* that establishes a truth-value cause-effect relationship between two segments whose propositional content can be tested in the real world (his being ill is *a fact* that can be objectively proved), and a *because* that relates an illocutionary act: a speaker's claim or conclusion (*He's ill*) and the justification or evidence he has for claiming so (he doesn't look well; the speaker can see it by looking at his face). The type of evidence is, in this case, sensory or perceptual (he *sees* that he is not well). Chafe suggests that sensory evidence indicates the kind of evidence on which inference is based (1986:267). The coherence of this second example is based on the world the text describes that, in this case, is a personal impression of the speaker when looking at the other's face; the state of affairs that is introduced by the second segment is not the cause of the state of affairs of the first segment, but the justification for making that claim.

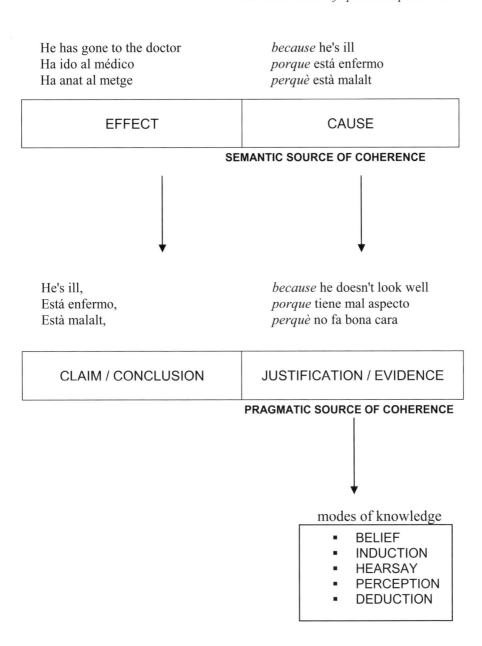

Figure 1. Causal-evidential connective *because/porque (Spanish)/perquè (Catalan)*

Chafe distinguishes five main types of *modes of knowledge,* each of which with its corresponding *source of knowledge*: belief, induction or inference, hearsay, sensory or perceptual evidence, and deduction.

Belief does not have a clear-cut source of knowledge. According to Chafe, "Belief is a mode of 'knowing' in which concern for evidence is downgraded. People believe things because other people whose views they respect believe them too, or simply because, for whatever reason, they want to believe them." (1986: 266). In the analysis carried out in the present study, we have considered "belief" those instances of *because* that introduce shared-knowledge-type of evidence, more specifically, knowledge that is accepted by a group that shares cultural and social rules. For instance, if at 11 pm a speaker claims "I think it is too late to call her at home now" or "I wouldn't call her now; it's too late," he's using an argument that is valid in most Western societies, i.e., calling someone at home late in the evening is not polite. Evidence plays a key role in *induction or inference,* although the nature of that evidence is usually not indicated (Chafe 1986: 266). In English, *must* is the most common marker of evidence, signaling a high degree of reliability, too: "It must have been a very difficult decision." *Hearsay* evidence is that which has been acquired through other people's experiences and has been told to us through language. It is the type of evidence from which the first studies on evidentiality stem, in Amerindian, African, and Asian minority languages. Typical examples would be storytelling and past experience accounts. Hearsay evidence is the most common since it also includes knowledge that speakers of a language acquire through readings (books, newspapers, encyclopedias, etc.) of other people's work. An interesting hearsay Catalan evidential is the form *es veu que* (literally, "one can see that", but semantically equivalent to "apparently"/ "It seems"), connected to visual sensory evidence and with a strong inferential value (González 2005: 531–33). As for *sensory* or *perceptual* evidence, Chafe (1986: 267) suggests that it is the kind of evidence on which inference is based. In English, speakers use the verbs *see, hear* and *feel*; when the reliability is lower, they adopt forms like *look like, sound like* and *feel like.* The verb *seem* is often found too: "She seems to be drunk" (listening to the way she talks). Finally, *deduction* predicts what will count as evidence by means of a reasoning process and hypotheses from which to deduce that evidence. The most common markers of deduction in English are verbal forms related to different degrees of reliability: *can* and *could* express a low degree of reliability; *should*, and *would* express hypothetical knowledge. The adverb *presumably* also falls within the deduction category (Chafe 1986: 269): "He shouldn't take so long;" "He writes the words to suit the tune, presumably."

The analysis of an oral corpus of eleven conversational texts on women's construction of genre subjectivity has shown that most relations established by Catalan causal connective *perquè* are pragmatic and that the justification or evidence that the interviewed women hold for making a claim often results in the creation of explicative utterances. In other words, what we traditionally refer to as 'cause' is, in the study carried out, the expression of pragmatic reasons that make the speaker's assertions socially acceptable; the cause becomes then a pragmatic evidence. The aim of the study has been to show that when speakers make use of pragmatic *perquè* the source and mode of knowledge are of primary importance, taking into account that causal relations originate from different sorts of evidence that speakers interpret. The reasons that people usually accept as valid somehow present *evidential* traits, that is to say, they mostly reflect our reasonings and our way to experience the world. Following Chafe's (1986) classification, we have observed that the coherence of the texts analyzed results from the different modes of knowledge that are linguistically expressed in the pragmatic reasons introduced by the causal connective *perquè*. Such observation has taken us to conclude that an opinion can be formulated as, for example, an induction or a belief, and that the notion of "reality" can therefore vary depending on the way the utterance is formulated.

However, the results have proved that Chafe's classification on *sources of knowledge* still requires going deeper into each sort of mode. The notion of "belief," understood as general shared knowledge, can be narrowed down and divided into *collective* beliefs, which are socially accepted (by a cultural community or a group of professionals, for instance), and *individual* beliefs and opinions, which are usually debatable and raise controversy. See both exemplified in (11) and (12) following:

(11) Collective belief
Volen ser dones que ocupen el mateix i fan exactament tot igual que ho faria un home. Segurament més bé, perquè les dones estem acostumades a fer moltes coses alhora, no? (PR)

They want to be women that occupy the same positions and do exactly the same as men do. Probably better because women are used to doing many things at once, you know.

(12) Individual belief
perquè justament és una autora que fa difícil que hi puguis estar d'acord; a vegades és profundament irritant (FB)

because she is a kind of author that makes it difficult that you can agree with her; sometimes she's extremely annoying.

Besides, the mode of knowledge that Chafe refers to as *perception*, that is, the sort of information that gets to us through the senses (by touching, hearing, seeing, smelling, and tasting), should also include what we call *emotional knowledge* which, in the corpus we have analyzed, is often linked to induction. See it exemplified in (13):

(13)
I aleshores no tant perquè em vaig sentir sola, sinó perquè ell ja no podia gaudir ni patir de la vida. (MAO)

And then not so much because I felt lonely, but because he could neither enjoy nor suffer from life.

Finally, in the same line of argument we consider that not all modes of knowledge have the same truth value for the interlocutor. Organized from more to less recurrent, we obtain the following ranking:

— *Induction.* Its value is undebatable since it is based on empiric observation and on the evidence of shared encyclopedic knowledge.
— *Beliefs.* Its value as a form of evidence is arguable, but the power that beliefs have in the building up of collective unconscious is indisputable. Being part of a culture or a group (ideological, professional, etc.) is basically a matter of shared beliefs.
— *Deduction.* Its value is indisputable if one accepts the premises. It is the most creative mode of knowledge for the many especulative possibilities that it offers. It has a great transformational power (we can imagine new worlds and deduct new forms of relation), but it can also have a manipulative side if we start from a fallacy.
— *Perception.* Its truth value depends on the sort of sense on which knowledge is based. Sight and tact are clearly perceived (*I've seen it with my eyes*), but taste and feelings aren't.

— *Hearsay*. Its value is absolutely debatable. What gets to us through somebody who has, at the same time, acquired the information through somebody else has little credit for an interlocutor. It is for this reason that *hearsay* mode of knowledge barely appears in our conversational corpus of analysis (just once).

Findings have proved that the Catalan *perquè* found in the interviews function as explicative utterances that introduce reasons that seem evident to our interlocutors. These reasons are based on observations of the world around them, on their opinions, and on their deductions, all of which allows them to speculate about new ways of thinking their subjectivity.

5.   The study: method and hypotheses

The analysis of Catalan *perquè* as evidential marker has been carried out in a spoken corpus of eleven conversational texts elicited in Catalan. The researcher has interviewed ten women that hold outstanding political, economic and cultural positions in the Catalan society[4]. They have responded to eleven questions, and the interviews have been elicited and recorded in audiovisual format, resulting into a series of oral samples of spontaneous discourse. See following the questions they were asked:

1. If I ask you who you are, what will you respond?
2. Is the fact of being a woman an identity trait that you consider relevant? Why?
3. What does being a woman mean to you? Do you feel comfortable with that identity?
4. Is there something such as 'a woman's experience'? And a way of thinking?
5. From your profession, how do you live the fact of being a woman?
6. What do you think about genre positive discrimination?
7. Is it important that women have access to powerful positions that have traditionally been occupied by men? How do you think 'feminine' power is carried out?
8. The main objective of the feminist vindication has to be equality or the acknowledgement of difference?
9. Now that we are going through the so-called *post* periods (poststructuralism, postmodernism, etc.), what do you think postfeminism should aim at?
10. Select two women that are especially significant to you. Why have you chosen them?

11. Explain a biographical event, something you remember with special interest and is important in your life.

The hypotheses held in the study are three. First, that besides logico-argumentative coherence relations, *perquè* establishes a large range of pragmatic relations in spontaneous oral discourse. Second, that the causal relations set up by *perquè* stem from evidence that the speaker interprets differently. Causal structures are fundamental in the construction of certain epistemic contexts or in the construction of knowledge. Third, in causal relations, the *source of knowledge* or *evidence* that grounds a claim has a direct influence on the construction of social systems of general beliefs and on the creation of personal attitudes in front of reality. These beliefs and attitudes tend to be accepted by speakers and society in general as truths that are often unquestioned.

6.   Discussion of results

The study of the connective *because* in the corpus of conversational texts on women's identity has been done taking Sander's (1997) notion of *source of coherence* and Chafe's (1986) classification of evidentiality as framework. Findings have shown that most instances of the connective facilitated a pragmatic relation between prior and following segments. We have seen that in pragmatic relations the speaker makes use of causal *because* to introduce a justification for his claim that provides the listener with the evidence the speaker has for making such claim ("He's ill, *because he doesn't look well*"> visual perceptual evidence). In the following lines we will discuss the evidential uses of Catalan *perquè* in the corpus and will try to show the way speakers use causal relations to construct a frame of beliefs and social attitudes that are commonly accepted as unquestionable and truthful. See the quantitative results of the analysis in Tables (2) and (3).

Table 2. Quantitative results of Catalan perquè in the corpus of interviews.

Interviews	Number of words	Number of perquè
AV	2.358	21
AMM	2.193	10
AS	2.356	22
FB	3.207	12
IC	3.286	17
IB	4.671	21
MAO	3.220	22
MT	4.375	21
PR	5.325	35
MS	2.723	27

Table 3. Mode of knowledge of causal-evidential p*erquè*

	Belief	Induction	Hearsay	Perception	Deduction	SSC	Final	Total
AV	5	11			5			21
AMM	1	8			1			10
AS	6	9			5		2	22
FB	3	7			1		1	12
IC	6	6			5			17
IB	5	8		3	4	1		21
MAO	2	11		5	2	1	1	22
MT	6	8			4	2	1	21
PR	12	22			1			35
MS	11	10	1	3	2			27
	57	**100**	**1**	**11**	**30**	**4**	**5**	**208**

The analysis of the corpus has shown that *induction* is the dominant mode of knowing and that *hearsay* is barely present in the answers provided by the interviewed women. Ifantidou (2001) based her work on evidentiality on the inferential processes that speakers make to sustain the evidence that justifies their claims. The inductive method of inferential processes provide speakers with a strong degree of reliability, basically through observation (inference is based on perceptual evidence, according to Chafe) and knowledge of the world. The analysis carried out has shown that speakers make assumptions about attitudes and beliefs that they consider are shared by hearers. As previously mentioned, the cause-effect relationships established (at the illocutionary level) may have important consequences in terms of generalized social beliefs and personal attitudes that stem from them. See (14) for an example of this type of evidence taken from the corpus.

(14) Induction
En el terreny de l'empresa, per exemple, la dona és la que incorpora la idea del grup. Així, les empreses nord-americanes més punteres han descobert ara que la dona és un filó en l'economia d'alta empresa, bàsicament, **perquè** incorpora aquestes idees del pacte i del grup, que són elements de la intel·ligència emocional. (PR)

*In the business area, for example, women are those that incorporate the idea of team. Therefore, in the most successful Nord American companies, there has been found that they are the keystone of the finances, basically **because** they include the idea of pacts and teamwork, elements of emotional intelligence.*

As for *hearsay* knowledge, we have seen that it is acquired indirectly, through other peoples' experiences, and that it has been transmitted to us through language. Early studies on evidentiality showed that some Amerindian, African and Asian minority languages include hearsay particles that qualify the source of knowledge. The source, in these languages, is based on oral tradition; the speaker reports an event that has been told to him/her by another member of the community. The hearsay evidence we find nowadays is based on written tradition, mostly. Speakers of a language receive large amounts of information from different mass media, newspapers and television as the most common ones. Encyclopedic knowledge acquired through the reading of books and digital media can also be considered hearsay evidence. This would possibly explain the outstanding presence of this sort of evidence in general, but also explains the lack of it in the building up of identity, a cognitive process that is highly acquired through collective and individual beliefs and through deduction. See (15), the only sample found in the corpus.

(15) Hearsay
Tu ara em recordaves quan vaig conduir el tren essent president de RENFE, i el cert és que es van produir dues reaccions: una, de la diguem-ne Espanya profunda, **perquè** mentre a Catalunya em deien: "molt bé, molt bé!", a la resta d'Espanya hi havia gent que em deia: "¡Ay, cuidado, cuidado!". (MS)

*You just reminded me of the time I drove the train while being president of RENFE, and it is true that it provoked two different reactions: one, let's say, from deep Spain, **because** while in Catalonia I was told: "very good, very good!", in the rest of Spain there were people who told me: "Be careful, be careful!"*

According to our findings, *Belief* is the second most common type of evidential introduced by the Catalan causal connective *perquè*. Chafe points out that this sort of evidential does not really need any overt marking and that we usually transmit belief just by stating a blunt remark that communicates it ("It is too late to call her at home," rather than "I think/believe that it is too late to call her at home"). However, when referring to topics of social interests, our view is that belief should be understood in *common ground* terms, as a conviction that is treated as truthful and valid by a large group of people that share social and cultural traits. See it illustrated in (16).

(16) Belief
Aleshores, penso que en el món de la política, com que entenc que la política és representació, i els que hi són han de treballar, doncs, pel bé comú de la gent que representen, no és dolent buscar una manera d'encoratjar que les dones hi siguin. **Perquè** les dones hi han de ser. (MS)

*Then, I think that in the political world, since I understand that politics implies representation, and that those who are in that world have to work, then, for the benefit of all the people they represent, it is not a bad idea to look for a way to encourage women to be present there. **Because** women have to be there.*

The third type of evidence mostly found in the interviews is *deduction*. The reasoning process that leaps to a hypothesis and conclusions from which evidence can be derived is rare in spontaneous discourse that expresses a personal opinion but it is still important when providing one's point of view. See it exemplified in (17).

(17) Deduction
La pel·lícula començava amb la Isadora de petita i la nena cremava el certificat de boda dels seus pares. Això em va impressionar molt **perquè** em va fer pensar que el matrimoni era una cosa terrible. (IC)

Perquè si una accepta ser una excepció, accepta la regla de la qual és excepció (FB).

*The film started with Isadora being a child and the little girl burned her parents' marriage certificate. That impressed me a lot **because** it made me think of marriage as something terrible.*

***Because** if one accepts being an exception, she accepts the rule out of which she's an exception.*

Perceptual evidence introduced by the causal connective analyzed has proved to be of little importance in this highly subjective conversational corpus. According to Chafe, the knowledge on which this sort of evidence is based is very reliable, since the source are the senses, basically *feel, see* and *hear*; forms that express a lower degree of reliability, like *feel like, seem like* and *look like*, are also part of this sort of evidence. However, the analysis undertaken shows that this type of evidence presents sometimes a blurred borderline with other types of evidence, since the notion of perception is not always based on first-hand physical evidence (the speaker seeing, feeling and hearing someone or something) but on information that he has acquired through collective beliefs or shared type of knowledge. See an instance in (18).

(18) Perception
No ho sé, tot i així era molt forta aquesta voluntat de no voler fer el que tocava. Potser **perquè** no veia felices les dones de casa.

*I don't know, it was very strong the willingness not to do what was supposed to be done. Perhaps **because** she didn't see the women living at home happy.*

Finally, as Table 3 above shows, there are just four instances of *semantic source of coherence* (SSC), in Sanders' terms. They show a direct cause-consequence type of relationship and are of the type shown in (19) following.

(19) Causal
El llenguatge és masclista, **perquè** està fet pels homes. (MAO)

*Language is sexist **because** it is made by men.*

7. Final remarks

The study of the corpus of interviews of women that occupy outstanding positions in the Catalan society has raised interesting points to consider and to elaborate further. We have seen that spontaneous discourse presents a high number of argumentative relations of cause and effect that are not logically sustained. The coherence of such relations has to be analyzed in pragmatic illocutionary terms: the first segment of the utterance presents a claim that is justified in the second segment, with the causal connective bridging the two. They are *dicto* relations that show the personal attitude and subjectivity of the speaker, opposite to those of *re*, which are sustained by objective propositional logic described in the real world. Chafe's classification on modes and source of knowledge requires, to our understanding, further thought; the classification results sometimes too fuzzy and presents overlapping of categories when it comes to analyze a corpus that deals with issues of political, cultural and social interest. The analysis of Catalan *perquè* confirms the first hypothesis of the study, that is, that besides logico-semantic relations, causal connectives set up a high number of pragmatic ones. The second and third hypotheses are related and have been confirmed, too. Causal relations are established through markers of evidence that the speaker interprets. They are fundamental in the construction of certain epistemic contextual spaces, that is, in the construction of social systems of beliefs and of personal attitudes in front of reality that are often adopted as unquestionable and truthful.

Notes

1. As Brinton (1996:29) states, there is a plethora of terms used: discourse con- nectives, connectors (argumentative, pragmatic, textual), particles, text organ- izers, modalisateurs, gambits, evincives, fillers, discourse operators, pop and push markers, cue phrase or clue word are the most common.. Terms are usu- ally attached to specific linguistic schools and currents, so for instance the terms argumentative connector, modalisateur and operator are mostly found in Argumentative Theory research (Ducrot 1983; Moeschler 1983, 1989; Lusher

1993, 1994; Roulet 1985, 1991, 1997) and the terms connective, cue phrase and clue word are common in Relevance Theory research and in those studies that take a cognitive approach to linguistic phenomena (Blakemore 1992; Grosz and Sidner 1986; Mann et al. 1992; Sanders 1997), but this is not clearly cut at all and a great number of linguists prefer to use general terms such as particle, text organizer or discourse marker. In all cases, the written and oral modes are taken into consideration, although it is well-known that European traditions tend to analyze more the written registers (rf. text linguistics) and Anglo-Saxon currents the spoken ones (rf. discourse analysis).

2. González (2004: chapter 3) offers a thorough discussion on the role of pragmatic markers in the production and interpretation of oral narrative discourse.
3. The difference between enunciat-enunciació (re and dicto, respectively), coming from French linguistic tradition, makes it more clear, to my understanding.
4. We have interviewed a minister of Catalonia's autonomous government, a cinema director, an editor, an economist, a biologist, a sculptress, a journalist, a mountaineer, a writer, and a philosophy professor.

References

Bakhtine, Mikael
   1977   Le marxisme et la philosophie du langage: essai d'application de la méthode sociologique en linguistique. Paris: Minuit.
Bateman, John, and Klaas Jan Rondhuis
   1997   Coherence relations: Towards a General Specification. Wilbert Spooren and Rodie Risselada (eds.), *Special Issue on Discourse Markers*: *Discourse Processes* 24: 3–49.
Beaugrande, Robert de and Wolfgang Dressler
   1981   *Introduction to Text Linguistics.* London: Longman.
Blakemore, Diane
   1989   The organization of discourse. In Newmeyer, Frederick W. (ed.). *Linguistics: The Cambridge survey. Volume IV: Language: The socio-cultural context.* Cambridge: Cambridge University Press.
   1994   Evidence and modality. In *The encyclopedia of language and linguistics,* Ronald Asher and James Simpson (eds.), 1183–1186. Oxford: Pergamon Press.

Briz, Antonio
    1994   Hacia un análisis argumentativo de un texto coloquial. La incidencia de los conectores pragmáticos. *Verba* 21, Universidad de Santiago de Compostela: 369–388.
    2001   *El español coloquial en la conversación.* Barcelona: Ariel.
Chafe, Wallace
    1986   Evidentiality in English conversation and academic writing. In *Evidentiality: The Linguistic Coding of Epistemology,* Wallace Chafe and Johanna Nichols (eds.), 261–272. Norwood, NJ: Ablex.
Chafe, Wallace and Johanna Nichols (eds.)
    1986   *Evidentiality: The Linguistic Coding of Epistemology.* Norwood, NJ: Ablex.
De Haan, Ferdinand
    1999   Evidentiality and epistemic modality: Setting boundaries. *Southwest Journal of Linguistics* 18: 83–101.
Dijk, Teun A.van
    1977   *Text and context. Explorations in the Semantics and Pragmatics of Discourse.* London: Longman.
Ducrot, Oswald
    1983   Operateurs argumentatifs et visée argumentative. *Cahiers de Linguistique* 5: 7–36.
Fuentes, Catalina
    1996   *La sintaxis de los relacionantes supraoracionales.* Madrid: Arco Libros.
González, Montserrat
    2004   *Pragmatic Markers in Oral Narrative. The Case of English and Catalan* [*Pragmatics & Beyond* NS 122]. Amsterdam/Philadelphia: John Benjamins.
    2005   An Approach to Catalan Evidentiality. *Intercultural Pragmatics* 2 (4). Berlin/New York: Mouton de Gruyter: 515–540
Halliday, Michael A.K. and Ruqaiya Hasan
    1976   *Cohesion in English.* London: Longman.
Ifantidou, Elly
    2001   *Evidentials and Relevance.* Pragmatics & Beyond New Series 86. Amsterdam/Philadelphia: John Benjamins.
Mann, William, and Sandra Thompson
    1988   Rhetorical Structure Theory: toward a functional theory of text organization. *Text* 8: 243–281.
Marín-Arrese, Juana I. (ed.)
    2004   *Perspectives on Evidentiality and Modality.* Madrid: Editorial Complutense.

Martin, James R.
　1992　　*English Text. System and Structure.* Amsterdam/Philadelphia: John
　　　　　Benjamins.
Plungian, Vladimir A.
　2001　　The place of evidentiality within the universal grammatical space.
　　　　　Journal of Pragmatics 33: 349–357.
Redeker, Gisela
　1990　　Ideational and Pragmatic Markers of Discourse Structure. *Journal of*
　　　　　Pragmatics 14: 367–381.
Sanders, Ted, Wilbert Spooren, and Leo Noordman
　1993　　Coherence Relations in a Cognitive Theory of Discourse
　　　　　Representation. *Cognitive Linguistics* 4 (2): 93–133.
Sanders, Ted
　1997　　Semantic and Pragmatic Sources of Coherence: On the Categorization
　　　　　of Coherence Relations in Context. Wilbert Spooren and Rodie
　　　　　Risselada (eds.), *Special Issue on Discourse Markers: Discourse*
　　　　　Processes 24:119–147.
Sweetser, Eve
　1990　　*From etymology to pragmatics. Metaphorical and cultural aspects of*
　　　　　semantic structure. Cambridge: Cambridge University Press.
Viana, Amadeu and Jordi Suïls
　2002　　Les construccions causals i les finals. *Gramàtica del català*
　　　　　contemporani. Volum 3, Sintaxi, capítol 27: 2937–2975.
Wierzbicka, Anna
　1994　　Semantics and epistemology: the meaning of 'evidentials' in a cross-
　　　　　linguistic perspective. *Language Sciences* 16(1): 81–137.
Willett, Thomas
　1988　　A cross-linguistic survey of the grammaticization of evidentiality.
　　　　　Studies in Language 12(1): 51–97.

A new look at common ground: memory, egocentrism, and joint meaning

Herbert L. Colston

1. Introduction

In a recent book edited by Colston and Katz (2005), two chapters were presented that, in their adjacency, offer a succinct demonstration of a current recent divergence in theorizing about contextual influences on language use and comprehension—that concerning the role of mutual knowledge in production and comprehension. The chapter presented by Richard Gerrig and William Horton (Gerrig and Horton 2005) revealed the benefit of applying traditional accounts of mutual knowledge on explaining the use and processing of contextual expressions (e.g., saying, "Seinfeld never had a jump-the-shark episode"). Contrasting with this view was the evidence summarized in the chapter by Dale Barr and Boaz Keysar (Barr and Keysar 2005) that showed egocentric language use being far more prevalent than could be accounted for by traditional accounts. A very brief synopsis and some speculation on this development were also provided (Colston 2005: 3).

> This is a fascinating development in that such contrary claims with evidence can be provided. My suspicion is that both claims are in a sense true, in that common ground is used, on occasion necessarily so, for important aspects of comprehension of many language forms. But there may also exist parallel lower level interaction mechanisms, as described by Barr and Keysar, as well as some others (e.g., mimicry, response in kind, script adherence, priming, chaining, attitude display, acting, and mere continuance) that are also influential. Future work might attempt to substantiate the separation of these mechanisms and address related questions such as how the mechanisms work individually and interactively, their possible interdependence, when are each used, and so on.

The following chapter will revisit and expand on this issue of mutual knowledge and its role in language use and comprehension. One goal of the chapter is to explore possible means of reconciliation in this separation between accounts of mutual knowledge. I will discuss a range of possible ways that both egocentric and exocentric language use and comprehension might coexist. Prior to that, however, I will first provide a treatment of common ground as memory, to point out how common ground as it is currently theoretically configured may need revision. Next, I will discuss an array of mechanisms that may, at least in part, guide, influence, dissuade, or marshal what speakers say or how they say it, that may intricately interact with rules of relevance or grounding without being strictly governed by them. I will then conclude with a final discussion of mechanisms that may allow shared meaning to arise that don't require strict common ground adherence. In this final discussion I raise the possibility of looking at common ground and relevance in a new way that might also help alleviate seeming disparities in the current accounts that deal with mutual knowledge consideration in language. Let me begin first though by very briefly explicating the nature of the diverging accounts.

2.   Common ground and egocentrism

Pragmatic theories make varyingly strict claims about the degree to which speakers use what they and their audiences and interlocutors know (and mutually know) to guide what is said and comprehended. Traditional accounts like Relevance Theory (Gibbs and Tendahl 2006; Carston and Powell 2006; Sperber 2005), and Herb Clark and colleagues' view of language as joint activity (Bangerter, Clark, and Katz 2004; Bangerter and Clark 2003; Clark, Schreuder, and Buttrick 1983), for instance, claim that successful language use and comprehension are guided, and indeed enabled, by consideration of knowledge and its mutuality. This view was nicely summarized by the organizers of this volume (Kecskes and Mey) in their letter to contributors:

> Traditional pragmatic theories thus emphasize the importance of intention, cooperation, common ground, mutual knowledge, relevance, and commitment in executing communicative acts. Stalnaker (1978:321) argued that 'it is part of the concept of presupposition that the speaker assumes that the members of his audience presuppose everything that he presupposes.' Jaszczolt's default semantics is based on the assumption that semantic representation is established with the help of intentions in communication. According to Mey's (2001) Communicative Principle, intention, cooperation and relevance are all responsible for communication action in a concrete context. Gregory, Healy and Jurafsky's (2002) results demonstrated that mutual knowledge is an important factor in production.

Common ground, relevance, and other similar hinge pins of the traditional accounts do seem to clearly dictate much or arguably even most of what a speaker chooses to say in a variety of conversational contexts, as well as how that talk is comprehended, particularly in situations where joint activity is crucial for a task at hand. The rules governing their maintenance may not, however, be broad enough an umbrella to account for all of what speakers say or comprehend in all circumstances. Other recent evidence suggests that some successful language use and comprehension can arise from more egocentric language behavior where interlocutors don't necessarily, or at least fully, corral their language production and comprehension with mutual knowledge consideration. Building upon evidence that speakers behave more egocentrically than allowed for by traditional pragmatic theories, this newer view has thus been placed at odds with the tenets of older accounts. Again, I'll simply quote the letter from the volume organizers, which described this view quite succinctly:

> For instance, recent research in cognitive psychology, linguistic pragmatics, and intercultural communication has directed attention to issues that warrant some revision of the major tenets of the traditional accounts. Several researchers (e.g. Stalnaker 1978; Keysar and Bly 1995; Barr and Keysar 2005; Giora 2003) indicated that speakers and listeners are egocentric to a surprising degree, and individual, egocentric endeavors of interactants play a much more decisive role in communication than current pragmatic theories envision. Investigating intercultural communication, Kecskes (2007) argued that instead of looking for common ground, lingua franca speakers articulated their own thoughts with linguistic means that they could easily use. Barr and Keysar (2005) claimed that speakers and listeners commonly violate their mutual knowledge when they produce and understand language. Their behavior is egocentric because it is rooted in the speakers' or listeners' own knowledge instead of in mutual knowledge.

There is thus a fairly strong divergence in theorizing about the role of mutual knowledge in language use and comprehension. The view presented here follows from the brief treatment in the Colston and Katz book (2005), arguing that there may be room for both kinds of account. To lay the groundwork for this discussion, let me first delineate how a consideration of common ground as a form of human memory may require a reanalysis of the role common ground plays in language use and comprehension.

3.   Common ground as memory

One of the most important insights, in my view, of the Horton and Gerrig work on common ground (Horton and Gerrig 2005), was its illustration that common ground is essentially a form of human memory. Rather than just being a theoretical construct about mutually known knowledge among interlocutors that supports and corrals language use and comprehension, common ground is additionally information that an interlocutor generates or encounters and then encodes into short- and/or long-term memory for ongoing or later use.

What this recognition unleashes, however, is the need for a much more thorough treatment of common ground as memory, including the entire set of reliable phenomena concerning human memory well established since the earliest days of the psychological study of memory. This treatment will need to include how memory succeeds and, perhaps more importantly, how it fails. Consideration of these phenomena may require a rethinking of common ground's role in language use and comprehension.

What follows in this section will not provide this thorough treatment. However, it will at least discuss some of what will need to be incorporated in further development of common ground as a theoretical framework that guides language use and comprehension. A number of these phenomena will likely be known to a psychological readership familiar with memory research, but its application to language use and comprehension anchored on common ground considerations may be somewhat more novel, particularly so to a broader audience.

3.1. Primacy and recency

A very well established effect in human memory holds that, all else held equal, memory for information that was encoded at the very beginning or very end of some study set, learning window, or other stream of serially presented information, will be greater than memory for information encountered in the middle of those sets (Murdoch 1962). If a person reads a long list of shopping list items for instance, later recall will be best for the items nearest the beginning (the primacy effect) and end (the recency effect) of the list.

Applying this notion to common ground as memory, recall would likely be greatest, again all else held equal, for items most initially and most recently grounded in a conversation (or re-grounded through the process of recall of common ground). This could have significant implications for common ground theorizing where previously it was generally held that grounding was simply something that took place (or not) in a conversation, with less consideration given to the serial position in which some information was grounded (or re-grounded) among all the other information encountered. If some key piece of information (e.g., that an interlocutor is not available for Tuesday meetings), was grounded in the middle of some series of other pieces of information that were grounded nearer the beginning or end of a conversation (e.g., that the interlocutor wants to participate in a meeting and has some important points to make), it may not be recalled. This could lead to misunderstandings (e.g., "Okay then, we'll meet next Tuesday"), along with the pragmatic implications of those misunderstandings (e.g., annoyance), and the subsequent need for repair ("I just said that I'm not free on Tuesdays!". "Oh that's right, you can't meet Tuesdays; we'll make it Wednesday").

It could even be possible that primacy and recency have an effect on the general structure of conversational grounding. For instance, is there a general structure to conversations such that they are more likely to have important information grounded near their beginnings and ends because of this characteristic of human memory (e.g., getting in the last word)? One could also pose similar questions for the other phenomena concerning human memory discussed here.

3.2. Anchoring effects

A great deal of memory research has also looked at effects such as anchoring. To demonstrate, consider the following classic anchoring task. A group of randomly selected U.S. residents is asked to guess the length of the Mississippi River. Prior to hearing this question, though, they are first asked to answer a yes/no question about the length of the river. Half of the people (randomly chosen) are asked, "Is the Mississippi River greater or less than 500 miles in length". The other half of the people are asked, "Is the Mississippi River greater or less than 5,000 miles in length". The results are compelling. The group asked the "500 miles" question, reliably gives an average answer that is much smaller than the group asked the "5,000 miles" question, even though both groups are equally accurate in answering those initial questions. Also, given random assignment, the groups start out equally knowledgeable about the topic at hand, so something about the anchoring question itself has to cause the difference in estimates. The explanation for this difference, although differing in detail depending upon the specific model being discussed, is essentially that people become anchored on the initial quantity they hear and then must make their subsequent estimate in light of having that initial quantity in mind. They tend to not deviate drastically from that initial anchor, such that the two groups end up giving different answers.

Applied to a common ground situation, a group of interlocutors could ground some initial quantity, polarity or magnitude at, for whatever reason, a particular level (e.g., Juan and Jean agree in explaining to Hasmig that the house party the two of them attended was "enormous"). Subsequent attempts at specifying the size of the party could then become anchored around "enormous," such that the interlocutors negotiate and ground the estimated size at 90 people. Had the initial grounding of the size used the description "tiny" instead, then the subsequent shared estimate could be quite different, perhaps 40 people. So what gets subsequently grounded (the number of people that Hasmig, Juan, and Jean believe that they know were at the party), can be greatly affected by what got initially grounded (the party was "enormous" versus "tiny").

3.3. Contrast and assimilation effects

The anchoring effect has been considered by some researchers as a specific instance of a broader effect known as assimilation. Assimilation effects are also often discussed along with their sister effect known as contrast, so as is usually the case, I'll treat them here together.

To demonstrate these effects, consider another classic memory task that shows contrast and assimilation effects. Three groups of people are provided with some objective piece of information. For example, they watch a short film that depicts an old, rusty car in poor condition that approaches the camera and then stops. A driver then turns off the engine, gets out of the car (the car door's hinges can be heard to screech), and then grabs the car door with both hands and slams it shut, hard. The driver then walks off camera and the film ends.

All three groups of people are then later asked to recall this film, and to make some ratings as measures of their representations of the event. But before doing this recall, the groups first have to do some intervening tasks.

One group, the control, is just asked to do some unrelated tasks that occupy them for a period of time. A second group, the assimilation group, is asked to consider a short list of well-known historical or fictional characters who all share the characteristic of being slightly dangerous, aggressive, mean or violent. What is specifically done with these characters is not of great importance, so long as the study participants consider those characters for a period of time and attend to the shared traits of the characters. The third, contrast, group does the same thing as the assimilation group, only the characters for the contrast group are well known for being *extremely* dangerous, aggressive, mean or violent.

Finally, all groups are asked to recall the film and rate some characteristics of the person shown on the film. Of particular interest is a question asking the participants to rate how violent they think the person is who, again, is shown slamming a car door. The control group typically provides a rating somewhere in the middle of the scale, which usually ranges from, "not at all violent" to "extremely violent", or something similar. Some participants in that group might view the door-slamming as indicative of a violent personality where others may see it a necessity of an old car whose doors don't easily latch, resulting in an average rating near the middle. Relative to the ratings of the control group, the assimilation group will often provide an average rating that is slightly nearer the violent of the scale. The average rating of the contrast group, however, will typically be nearer the non-violent of the scale relative to the control group, with a reliably greater magnitude of difference from the control group's average rating compared to the difference between the assimilation and control groups.

The explanations for these differences correspond to the mechanisms of assimilation and contrast. For the assimilation group, the exposure to moderately violent historical or fictional characters serves to activate the concept of moderately violent behavior, which is still active when the participants consider the film character's personality at the recall task (Herr 1986; Smith and Branscombe 1988; Ford, Stangor, and Duan 1994). This results in that character's behavior being interpreted as stemming from a moderately violent personality, rather than from some situational necessity, given the heightened availability of the concept of violence. For the contrast group a similar activation of a violence concept also takes place, but for this group since the activated concept is much more extreme, the violence of the film character pales in comparison. Relative to the violence of Adolph Hitler, John Wayne Gacy, or Freddy Krueger, the film character indeed seems much less violent and is rated accordingly.

Of course the magnitudes of difference between the activated information in the biasing contexts (the consideration of moderately or extremely violent historical/fictional characters) and the target judgment (how violent is the man in the film), will affect whether assimilation, contrast, or no biasing effects occur. Also the range and type of considerations being made (personality characteristics, magnitudes, physical characteristics, perceptual judgments, etc.) and the degree of juxtaposition of the biasing and target items are crucial. But the overall general pattern holds in a great number of different areas of judgment—biasing information that is slightly different from a target characteristic will typically pull judgments in the direction of the biasing information. Biasing information that more drastically differs from a target characteristic will typically push judgments away from the biasing information.

Applied to common ground, the typical pattern of contrast or assimilation effects would also likely hold. Initially grounded information can serve as a biasing context for later grounded information (and indeed, vice versa). If the initial information is slightly different or drastically different from subsequently grounded information, assimilation and contrast effects respectively can occur such that the encoded representation of the later grounded information may change compared to had it been encoded in isolation.

3.4. Advantage of first mention

Yet another common memory effect concerns the heightened recall for the initial piece of information encountered in a pair of pieces of information (e.g., two names). This "Advantage of first mention" has been demonstrated in multiple languages, and seems to not be due to syntactic, semantic, or other factors other than the mere order of presentation. The likely explanation here concerns the sequential nature of representation formation, such that first-mentioned information has greater recall accessibility because it forms the foundation for the broader sentence level representation, which includes the second-mentioned information (Gernsbacher and Hargreaves 1988; Gernsbacher, Hargreaves, and Beeman 1989; McDonald and Shaibe 2002; Carreiras, Gernsbacher, and Villa 1995; Smith et al. 2005; Jarvikivi et al., 2005; Kim, Lee, and Gernsbacher 2004).

Applied to common ground, the advantage of first mention could be a key influence in that information that is initially grounded in a conversation plays a much more central role in representation formation and is more likely recalled later, relative to subsequently grounded information.

3.5. Schematic knowledge

It has long been known that knowledge in memory has a schematic structure, such that encountering one small part of a schema (e.g., menus in a restaurant) actually seems to activate an entire set of related information (the entire restaurant script). This also can cause memory errors in that people will later "recall" information that hadn't actually been explicitly encountered before, but instead was simply part of an activated schema (Bartlett 1932).

This memory phenomenon could once again have a major influence on how common ground works in that information grounded in a conversation could easily trigger pre-existing schematic structures in memory. A speaker might then later think that some information is in the common ground with an interlocutor, when that information had not, in fact, been grounded with that interlocutor. Rather, the additional information was simply part of a schematic structure that got activated.

3.6. Congruency and memory

Another classic domain of memory has shown that, all else being equal, memory will be better if recall is done under the same context in which the information was originally encoded in memory. If the study and recall contexts differ, then memory will suffer.

This effect clearly impacts common ground in that people are always changing contexts, both cognitive and physical, that can then alter which parts of common ground are recalled better (or worse). If a speaker grounded some information when talking with two other people, for example, a new conversation taking place later with just one of those people might not be able to rely on the supposed common ground between those two interlocutors because of the change in social context. But if that later conversation were held between the three original people, then common ground might be more available.

3.7. False memories

Perhaps the most compelling of the memory effects that might impact common ground is the ability for false memories to be implanted in people without their knowing.   A very powerful mechanism of memory implantation has been studied in depth by Elizabeth Loftus and colleagues as a means of assessing the reliability of so-called repressed memories (Loftus and Cahill 2007). All that is really required for these implantations to happen is to present a plausible autobiographical event to a person under some realistic guise for discussing that past event. Then have the person recall genuine memories from the past that occurred at the time the false event was to have happened. Have the person repeatedly and over some period of time reconsider the false event alongside the actual past events and attempt to imagine them all happening. Over time the person will take in the false event and begin to "recall" it as if it actually happened.

This false memory implantation has an enormous impact on our view of common ground.   It demonstrates that what people consider actual memories might instead be what the people have negotiated those memories to be with an interlocutor, rather than any event that actually occurred. This idea is taken up in section 6 below, but suffice it for now to say that common ground may thus in some instances function backwards from what we've thus far thought. What a person currently thinks he or she and an interlocutor mutually know about some past event, supposedly because that event occurred and the people saw each other experience it and thus put it in common ground and now remember it, may instead be a *false belief* about that past event that gets created just because the interlocutors are now talking about it.

3.8. Memory and embodiment

The effects above stem from a view that human memory is a general cognitive capacity that, although having its sets of shortcomings, is nonetheless a fairly encapsulated ability of the mind that operates as it does independently of other physiological or biological processes. The content of what is being remembered itself is of little importance because it can all be reduced to bits of information that are encoded, stored, retrieved (or not), etc., from an all-purpose human memory system.

It turns out that this view, although in limited contexts can be reliable, as an overall explanation of memory is an oversimplification. Human memory, in a broad sense, is much more like other bio-cognitive processes in that it has been shaped through evolution to service the needs of the gene, individual, and social group. It will thus operate differently according to certain characteristics about what is being remembered and what those memories are for. Space limitations prevent a full outline of this idea, but it turns out that different kinds of memory systems may serve different needs, and thus operate more and less accurately depending upon the content and consequences of what is being recalled.

As a consequence for common ground, recall of mutually shared information might be much better with people in whom we have some kind of interest (e.g., sexual, social, etc.) compared to other people we find less important, even if the degree of initially grounded information between the pairs of people started out equivalent.

3.9. Summary

The effects treated here are by no means an exhaustive list of memory phenomena that might have an influence on how common ground operates to enable language use and comprehension. Among many other memory phenomena that space limitations prevent full discussion of here are; distinctiveness, familiarity, intentional versus incidental encoding, levels of processing, proactive and retroactive interference, context effects, priming, a wide variety of serial position effects, blocking, read versus generate recall differences, and decay and interference models of forgetting. A full accounting of how common ground works in language will require a thorough consideration of all of these.

The effects and examples that are included here though, also nicely demonstrate that what has already been well established by memory research–that human memory is highly malleable, dynamic, vulnerable to pre- and post-event information, schematically structured, variable according to motivation and content, and overall, demonstrably unreliable, is also true of common ground. But common ground as a form of shared memory may indeed multiply these uncertainties across all the interlocutors involved in a conversation. If I recall an event inaccurately because of my own biases or due to other information I've encountered (e.g., I describe a reasonably large birthday party with "the party was tiny" because I had recently been watching television footage of the enormous New Year's Eve party in Times Square, New York), that can cascade through my interlocutors such that they also believe the birthday party was "tiny." This can then subsequently affect all of our later cognition about the topic (e.g., we concur that there were few people at the party).

Of course one could also make the opposite argument that common ground, in being a form of shared memory, should be less vulnerable to such manipulability or malleability, given that more than one person is available to keep recall accurate. If one person initially describes a large party as "tiny," there is another person's memory available to check the potential inaccuracy of that description ("that party wasn't tiny, it was actually pretty big"). Certainly, this double-checking of accuracy is possible, and doubtless on occasion does occur. But one must also then attend to a great deal of social interaction phenomena that would undoubtedly play a role in what gets collectively recalled and established in interlocutors' common ground. Only in cases where people cling hard to their own subjective recall of events and then argue publicly in favor of their version of those events, could such a collective increase in accuracy be possible (but the accuracy increase would still require fairly accurate initial subjective recalls as well, which themselves aren't reliable). If the initial description ("tiny") instead were to come from a social authority figure of some kind, other people may defer to that description even if they initially disagree, and then memory and cognitive dissonance mechanisms can initiate and the collective memory can get changed to fit that authority figure's recall (Cuc et al. 2006; Weldon 2001).

Essentially, embodied, personality, social, cultural, and other factors can play an enormous role in affecting the output of so-called encapsulated cognitive processes involved in memory and language, far more that some objectivist oriented cognitive models of event representation and common ground would allow. Although the famous fable has one brave person pointing out the Emperor's missing clothes, in reality, many people will indeed later "remember" the garments being there. Future theorizing about how common ground gets used in language production and comprehension must address these factors.

In the next section I will turn to the variety of motivations for why speakers talk, and the degree to which the motivations may (or may not) involve common ground considerations. Even if we acknowledge from the evidence in the current section that common ground is a much less stable base for language use and comprehension than has been previously thought, it turns out that for many kinds of talk speakers may not greatly use it anyway.

4.　Why do people talk? Exocentric and egocentric mechanisms of production.

If one considers all the motivations for why people talk in all the wide variety of discourse contexts where talk occurs, a great many of these motivations may not demand strict adherence to the mutual knowledge between interlocutors for language production.　Motivations for talking certainly *could* be, and indeed often *are* constrained by such considerations, but they need not, and in reality often simply do not, *require* common ground tracking for successful production and comprehension.　What follows then is a list of some of these motivations.　These are presented in no particular order; rather they just provide an array of the many reasons for why people talk in normal everyday conversations.

4.1. Drift

For a great array of different reasons, speakers will often simply let their minds drift in the midst of a conversation.　They might originally be thinking about one thing and say something relevant to that topic.　The addressee might even address what was said in a reply.　But while the addressee is making that response the speaker's mind wanders and they'll blurt out something else, often varyingly irrelevant to the train of the conversations—either to their original utterance or to the intervening response by the addressee.

This kind of non-sequitur is viewed as a mistake in traditional kinds of production theories, because the speaker failed to take account of the common ground they share with the addressee.　Yet it may be nothing of the kind.　It could just be something a speaker does as part of a normal conversation in which multiple complex demands are in place on the speaker.　Drift could also serve some underlying characteristics of the speaker, such as their social relationship and/or attitude toward the addressee.　Speakers might thus have a mind drift because they have a short attention span or like to creatively follow trains of thought.　Or they might drift because that is their established pattern of talking with a particular addressee or about a particular topic.

The point for purposes here is that such a lack of audience design is not necessarily incomprehensible, or indeed even uncooperative. Rather, the speaker may just be attempting to continue the conversation as best as possible. But the speaker is still acknowledging and working around (or possibly, selfishly over-attending to) their own thoughts concerning the addressee or topic.

4.2. Avoidance

Avoidance is another similar talk motivation that can arise for many kinds of reasons. It can happen if somebody has received some bad news that they don't want to hear such as some bad medical news about themselves. Or it can arise if a speaker doesn't like engaging a particular addressee. The speaker might even use avoidance because they *can't* engage the addressee. For instance, they might be intimidated by the addressee and don't know how to approach him or her, so they keep changing the topic. Or they may not hold the addressee in high esteem and so don't want to engage them in a conversation.

Avoidance thus produces talk that, because of the very nature of what the speaker is doing, tends to avoid the content at hand. It produces talk that purposefully deviates from that topic, that tries to steer the addressee's attention away, that tries to change the topic, to put a more humorous tone on things, tries to make jokes or other things to avoid talking about the topic. This avoidant talk can also belittle, demean, make fun of, or try to lend lesser importance to the target topic, because something that has lesser importance requires lesser attention.

In terms of common ground the avoidance mechanism is particularly interesting. By its very nature, avoidance will lead to a speaker purposefully *not* using common ground material, and instead making use of other material. A person, for a variety of reasons, is simply not engaging directly in a give-and-take dialog with their interlocutor on common ground, and yet can still be readily understood.

4.3. Reluctance

A similar but perhaps weaker mechanism could have the speaker not actively avoiding discussing some common ground content with an addressee, rather a speaker is simply reluctant to engage common ground content with someone and speaks accordingly.

4.4. Emotional expressions

A very prevalent motivation for talking that often does not involve audience design is the verbal expression of some emotion. Indeed, this motivation is frequently depicted in fictional settings because of its comic effect.

Consider the following transcription of a well-known episode of the American television program, the Jerry Seinfeld Show[1]. The regular characters Elaine, Jerry, and George, along with George's girlfriend Nina, have flown to India for a wedding between Elaine's old friends Pinter and Sue Ellen. Earlier in the episode, it was revealed to George that his lifelong friend Jerry had recently slept with Nina. George has been grumbling about this the entire trip. The exchange below takes place at the wedding. The bride and groom and the entire wedding party, including Elaine who is in the wedding, are present in a crowded room, and Elaine steps over to talk briefly with her friends before the ceremony. At this point, Elaine has had enough of George's complaining:

ELAINE: Would you grow up, George?! What is the difference? Nina slept with him (Points to Jerry), he slept with me, I slept with Pinter. Nobody cares! It's all ancient history.

GEORGE: (Loud, so everyone at the wedding can hear) You slept with the groom?! (Everyone goes silent. George and Elaine both look sheepish. All eyes on them, especially Sue Ellen's).

("The Betrayal", originally aired 11/20/97)

4.5. "Freudian" and other "slips"

Speakers will also often succumb to vocalizing subconscious or suppressed thought content in their utterances, which can violate the consciously intended direction of common ground development. Indeed, it can occasionally thwart those conscious intentions, sometimes also to great comic effect.

Two examples taken from the Simpsons animated series amusingly demonstrate this mechanism[1]. In the first example Homer's father, Abe Simpson, has begun dating Homer's wife Marge's elderly mother. During this courtship, wealthy power plant owner Montgomery Burns steps in and appears to be stealing Marge's mother away from Abe, despite Mr. Burns' obnoxious character flaws. Marge is bemoaning this turn of events, and Homer attempts to agree:

MARGE: [Mr. Burns] is an awful, awful, awful man! I guess if he makes Mom happy, that's all that really matters
HOMER: That's right money. Your money's happiness is all that moneys.

("Lady Bouvier's Lover", originally aired 5/12/94)

In the second example, Homer is speaking to the viewer in a behind-the-scenes of the show format as if the Simpsons were a real family that had their own show. He and Marge are discussing the difficulties of raising children,

MARGE: Nobody told us how tough it is to raise kids. They almost drove me to fortified wine.

HOMER: Then we figured out we could park them in front of the TV. That's how I was raised, and I turned out TV.

("Behind the Laughter", originally aired, 5/21/2000)

4.6. Responding in kind

Another talk motivation seems to arise from both a phatic unfolding of a conversational script and a kind of priming that occurs in an addressee's response to an initial utterance. In either case, a speaker essentially responds in kind to what they've just heard said by another speaker. In these responses, the speaker does not seem to be using audience design in the usual sense. Although it might be argued that the speaker is just adopting the easiest and safest form of audience design (e.g., just recycle the remark—if it was proper to use it addressing me then it is proper to use it in return). Responses in kind often appear in short small-talk exchanges, (e.g., "nice to meet you too," in response to hearing "nice to meet you").

4.7. Plagiarism

Speakers will also on occasion borrow precise segments of talk of other people's creation and portray them as their own. This can be done with common ground in mind, as in choosing a particular plagiarized segment that will enhance understanding in the addressee by working with what they already know. Or it can be done more for the speakers' purposes of trying to alter their internal representation of themselves.

4.8. Mimicry

A similar but broader mechanism than plagiarism is mimicry. Similar to the case of plagiarism, mimicry also can clearly involve common ground. A speaker may choose which aspects of someone they admire to mimic in speaking to other people, in part from an audience design consideration (e.g., don't mimic an admiree to an addressee who knows both you and the admiree; they'll likely notice your mimicry and possibly call you on it). But the mimicked talk can also be for more purely self-oriented purposes. A speaker admires another person and wishes to be like him or her. So she unconsciously or consciously copies that person's type of talk as a means of feeling that she is indeed similar to that admired person.

Young people in particular, who are still undergoing identity formation, will often experiment with borrowing different mannerisms or similar characteristics of other people, perhaps people they admire including how they talk, and will adopt them as their own. The focus of this activity is often mainly on getting those mannerisms right or simply enjoying doing those mannerisms, which can then come at the expense of full negotiation of common ground with an interlocutor.

4.9. Attitude expression

Oftentimes a speaker will be more motivated to express their private attitude about some topic rather than to get this attitude comprehended by an addressee or other hearer/reader. Particularly in the cases of very strong or negative attitudes, speakers often cannot help but express those feelings, indeed, even if they're trying to hide them. Certainly comprehenders can register these attitudes as well, but again not necessarily because of common ground. Rather, comprehenders in some cases are particularly attuned to detecting negative attitudes. In other cases, the very means by which a bit of language captures a negative attitude for a speaker also reveals it to a comprehender without common ground playing a role in the meaning exchange. As a brief example, asyndeton, a kind of language structure where all but the most crucial words are stripped from an utterance (e.g., I went, I ate, I left) can capture a negative attitude merely in its minimalist structure. A speaker with a negative attitude toward some topic may not wish to devote much time or effort in referring to that topic. This attitude then goes into the minimalist asyndeton structure and hearers can then see it when spoken.

4.10. Filibuster

People will also on occasion use language primarily for floor-holding reasons, rather than informative or interactive purposes. One last Simpson's example will be used to illustrate this motivation[1]. The filibuster motivation is observed regularly in the Grampa Abe Simpson character, who often uses lengthy rambling soliloquy's for this purpose. In the following example, Montgomery Burns' power plant employees are on strike, so Mr. Burns asks his assistant Mr. Smithers for some tough, old-fashioned strike breakers like those used in the 1930s to be brought in. Abe Simpson and some other elderly pals then appear:

> GRAMPA: We can't bust heads like we used to, but we have our ways.
> One trick is to tell 'em stories that don't go anywhere - like the time I
> caught the ferry over to Shelbyville. I needed a new heel for my shoe, so, I
> decided to go to Morganville, which is what they called Shelbyville in those
> days.

So I tied an onion to my belt, which was the style at the time. Now, to take
the ferry cost a nickel, and in those days, nickels had pictures of
bumblebees on 'em. 'Give me five bees for a quarter,' you'd say.
Now where were we? Oh yeah - the important thing was I had an onion on
my belt, which was the style at the time. They didn't have white onions
because of the war. The only
 thing you could get was those big yellow ones...

("Last Exit to Springfield", originally aired 3/11/1993)

4.11. Lubrication

Another motivation for a speaker to talk might be just to get another
person(s) to begin or continue talking. The original speaker could use
common ground in this attempt, which can be particularly effective by
tapping into content that the addressee knows the speaker knows about.
But again, use of common ground is not necessary in this context, and in
fact, not using common ground can be one way of getting the other person
to talk—by forcing them to clarify or correct something that was
incorrectly asserted. Or the original speaker could just prod the other
person to talk without any use or misuse of common ground.

4.12. Display

Still another motivation involves talk simply to dazzle or impress
interlocutors or audience members. This motivation also need not involve
tracking common ground. If it does, it could involve a speaker monitoring
what an addressee does not know, and then using words that are beyond
that capability to impress or dazzle. But the motivation could also be
revealed in a speaker just using the biggest words they know without caring
whether the addressee knows them or not.

4.13. Getting it down

Another talk mechanism that very poignantly ignores common ground
happens often when a person is composing a new idea aloud in the presence
of an interlocutor. Here a speaker will attempt to utter some bit of meaning

simply to anchor it for his or her own purposes. By stating the idea aloud, the speaker is working with a form of self-common ground, in that the external hearing of the idea places it more firmly in his or her memory. Such statements can then free up working memory to address new bits of meaning, so that thinking can progress. The very nature of this kind of talk frequently ignores the mutually known information with the interlocutor to instead focus on the formulation if a new idea by anchoring parts of the idea through speaking aloud.

4.14. Mere continuance

A form of talk motivation that is very similar to lubrication and responding in kind is mere continuance. With mere continuance a speaker will simply utter any bit of language just to have the conversation continue. This talk needn't prod the addressee to talk further, nor involve a repetition of something just said. Rather it just fills the silence after the other speaker's turn. This kind of talk is probably the least successful of all the other motivations that don't depend upon common ground if the interlocutors know one another very well. In those conversations there might be a high expectation of relevance. However, it is commonly used among strangers who simply have little common ground to work with but are nonetheless in a position to converse. This motivation allows them to talk without common ground.

4.15. Alignment

There may also be instances of talk where a person is demonstrating their alignment with something. For instance, some speakers will cite little pieces of popular culture, perhaps snippets from a television show or popular song, to display something about themselves. This of course can involve common ground, but it could also be just a form of showing off something about the speaker's identity.

Additional talk motivations that also might not necessitate common ground monitoring for production could also involve; phatic talk, interrogation, mockery, acting, pretense, script adherence, priming, questioning, quizzing, and story telling among many other possibilities. Full discussion of all of these is beyond the scope of the present work, but attending to these motivations might be warranted given they might reveal additional requirements of a revised view of common ground.

5.   How do people comprehend? Exocentrism and egocentrism in comprehension

The preceding section attempted to demonstrate a variety of occasions where speakers do not seem greatly concerned with audience design in the creation of their utterances.   This is by no means to say that the idea of common ground is somehow false or that common ground is never used in audience design or in comprehension—clearly, common ground plays a key role in both processes much of the time.   So, should the above examples and the great degree of egocentrism shown in recent research on speaker's language use just be seen as talk that deviates from successful common ground usage?   This could indeed explain why many of the above categories of talk are humorous (to the extent that deviance underlies humor).   However, if the above instances are not seen as simply mistakes in failing to use common ground, then what role does common ground play in comprehension? Must it always be used for successful comprehension?

I'd like to argue that, although there is some degree of deviance from common ground usage in some of the above instances of production, which probably does account for some of the humor involved, this is not the entire story.   There are very likely other mechanisms at play that allow for production and comprehension to proceed relatively smoothly in interlocutors without common ground playing a necessary role.

5.1. Drift and re-anchor

One such mechanism is simply to put the consideration of common ground into a holding pattern and save its role for much later in the comprehension process.   This drift and re-anchor mechanism could allow for the kinds of production drifts discussed above, where a speaker seems to utter non sequiturs in the midst of a conversation. All interlocutors need do is simply continue talking, and then at some point later, return directly or indirectly to the ambiguous statements and comprehend them at that point when more information is available.

One could argue that this mechanism is simply negotiation spread out over time, which in some ways it is. But it also can allow un-negotiated bits of meaning to simply fade away and never be grounded, without having them stall the other exchanges of meaning. Indeed, often speakers do not fully understand some point they are trying to make themselves, or have great difficulty in putting those points clearly into words, such that full common ground on those points isn't possible to begin with. These needn't, however, hold up other exchanges of meaning.

More global mechanisms can also overshadow these local ungrounded meanings such that the comprehenders may gain the illusion that grounding actually did happen. Indeed, oftentimes a level of agreement can be reached between individuals in some heated debate simply by forcing them to talk for some time. The cognitive dissonance involved in their having worked on some bits of meaning, can spread to other lesser understood bits of meaning such that the interlocutors walk away thinking they share more common ground than they possibly do.

There may also be an interesting interplay between tightly negotiated instances of common ground in a conversation and the drift-and-re-anchor mechanism. Interlocutors might entertain some degree of lack of understanding for a while until some threshold is reached, and then actively work to ground some of those ambiguities before proceeding.

5.2. Chase

Another mechanism involves the burden of comprehension being shifted nearly completely to one interlocutor. This chase mechanism, which can often arise in interlocutors with uneven social power, involves the less powerful member having to do extra comprehension work to keep up with the speaker who is not attending to common ground. Often the mere recognition of such an imbalance on the comprehender's part itself is a mechanism of comprehension. The comprehender simply allows that the speaker is not using common ground in their talk and is instead behaving egocentrically, and then attempts to use external contextual and other supports more extensively to aid comprehension.

5.3. Verbal play

Another mechanism challenges the idea that interlocutors are always seeking to exchange meaning. Many instances of talk and comprehension instead involve people simply engaging in varieties of verbal play with one another, for a large number of reasons, including simply play for the sake of play. Comprehension here is a much more open process that can but needn't have much to do with common ground.

5.4. Offloading

Other mechanisms can involve comprehenders pretending to have understood something (indeed, even to themselves), and then genuinely comprehending later, or at least allowing time to give them the illusion they've understood later.

5.5. Good enough comprehension

The notion that comprehension is some well agreed-upon, all-encompassing thing that happens or not, or even that happens partially, may require rethinking when we consider common ground and the role it plays in comprehension. A better description may be that comprehension is a loose continuum that ranges from something minimal to something richly elaborated upon with inferences, etc. This view would allow for a "good enough comprehension" for purposes at hand, than might align better with instances of talk that aren't wildly relevant for hearers or that don't make great use of common ground.

5.6. Resource allocation

Many of the discussions of common ground, including the new work that has noted the surprisingly prevalent egocentric nature of talk, has discussed the role that information processing demands play in common ground usage in production (and comprehension). If a speaker is overburdened with production processing demands, for instance, he or she is less likely to have resources to allocate for common ground consideration. I would only expand this view to include resource allocation among idea formation, language production, social considerations, emotional states, other cognitive considerations (e.g., working memory capacity), as well as strategic and other planning influences on production. Speakers may, for instance, have enough processing capacity during some language production act to consider their common ground with the addressee/audience, but their degree of comfort, their emotional state, the degree to which what they say now will matter later, among other things, can all influence the extent to which they use common ground in production (and comprehension).

5.7. Intentionality and common ground violation

Another issue that can affect common ground and its use in comprehension is the intentionality of a speaker's violation of using it. Some speakers may simply violate common ground in production because of the memory limitations discussed earlier. A speaker simply fails to retain some information in common ground, or has had his or her common ground representation undergo some alteration, such that a production by that speaker does not match common ground. This speaker does, though, *believe* that he or she is following common ground. Other speakers, however, may intentionally not worry about common ground in their productions, perhaps because of processing limitations, social expectations or other reasons. Of course, still other speakers could intentionally misuse common ground for other purposes (e.g., an interrogator trying to force a person to say they think something). These differences might in turn affect how productions are comprehended.

5.8. Culture and common ground

One other issue could be social or cultural differences in people's likelihood to attribute nonsense or non-consideration of common ground on a speaker's part. It could be, for instance that some people, as a fairly regular pattern, generally expect relevance from speakers. These comprehenders would then be in a difficult position if relevance is lacking in a production. Other people might be far more flexible in their expectations of production relevance. Whether this flexibility would arise from experience, from social interactions, from cultural differences, from regional differences in ways people talk, personality differences or even the requirements of different kinds of interaction, it might produce a reduced expectation of relevance in speakers' productions. These differences could also clearly impact the role common ground plays in comprehension.

5.9. Inevitability of egocentrism

It is also the case that speakers are to some degree always egocentric, or at the very least, they are not able to be completely non-egocentric. People can never know with complete precision what another person is thinking or knowing at a given moment, so it is inherently impossible to perfectly craft each and every utterance to be perfectly relevant for all hearers at all times. Some degree of approximation is always present and people will have to work with their own set of internal influences. The point for present purposes is that that degree of approximation nonetheless may vary, in that for some interlocutors at some times it is fairly minimal—as when interlocutors are very tightly maintaining common ground, monitoring it, and closely using it for production and comprehension.

5.10. Failure to ground

Lastly, consider the simple case of interlocutors failing to ground in conversations, yet those conversations continuing regardless. On these occasions, interlocutors will simply fail to ground some key bit of information or even repeatedly fail to ground such information but still make conversational contributions and comprehensions, such that the ensuing conversation, although possibly going somewhat awry, continues. This of course shows the importance of grounding for some kinds of talk, but it also shows that conversations can still happen and continue without grounding. Consider the pattern of exchanges in the following excerpt from Richard Russo's (1988) novel *The Risk Pool,* particularly the unnamed character introduced as "somebody" in the background, and the person who refers to his wife.

   Sam Hall, father of the teenage narrator Ned Hall, comes into a diner where his son and several other men are waiting and speaks to his son:

"You know that Schwartz kid?" he asked me one afternoon after our ritual greeting.

"Claude?" I said. I'd neither seen nor thought of him since school got out.

"His old man runs the factory out in Meco?"

I said that was the one.

"Tried to commit suicide this afternoon," he said. "Hung himself, the crazy son of a bitch."

A lunatic discussion ensued. Several people in the diner had heard of the event, or overheard somebody talking about it, just as they'd overheard my father's mention of it to me.

"Schwartz," somebody said. "Bernie Schwartz?"

"Bernie Schwartz is older than you. This was some kid."

"Maybe it was Bernie's kid," the original speaker suggested.

"Bernie never had no kids and he never run no factory in Meco. Other than that, it could have been Bernie."

Everybody laughed.

"It was Clyde Schwartz," my father said, getting it wrong, but close. "Third Avenue they live, somewhere."

"There's no Jews on Third Avenue. My wife lives up on Third Avenue."

"It's Clyde Schwartz," my father insisted. "And they live on Third Avenue, I'm telling you."

"What's he want to kill himself for if he owns a factory?"

"It's not him, it's his kid. Clean your ears."

"The Schwartzes live on Division Street, all of them. Right by the west entrance to the park. Except for Randy over on Mill."

The door opened and Skinny shuffled in, filthy and smelling of fertilizer from an afternoon in the Monsignor's flower beds.

"Hey, Skineet" my father hailed him. "Where does Clyde Schwartz live?"

"Third Avenue," Skinny said, happy to be deferred to in this local matter. "He damn near cooked his own goose today."

"Not him," my father said. "His kid."

"No, him is what I heard. Tried to string himself up from the ramada in his backyard."

"From the what?"

"I heard it was the kid," my father said, unsure of himself now.

"Couldn't be," Skinny said. "He tied a rope to the roof and jumped off the picnic table. Neighbor looked out the window and saw him standing there on his tiptoes, eyes all bugged out. When he didn't wave back, she got suspicious. Old Lady Agajanian."

"There's no Agajanian on Third Avenue," said the man whose wife lived there.

"Old Goddamn Lady Agajanian," Skinny shouted, "you simple shit! On Third Avenue. Next to Claude Goddamn Shwartz."

"Besides," somebody said. "Your wife lives on Second Avenue."

The man had to admit this was true. He'd forgot. His wife did live on Second Avenue.

"I heard it was the kid," my father said.

"All right," Skinny said. "You tell me how a kid's gonna bend down the crossbeam on that ramada."

"I'm just telling you what I heard," my father said, throwing up his hands. "Some kid named Clyde Schwartz tried to kill himself is what I heard. Sue me."

"I don't want to sue you. But I'll buy your dinner if you're right."

"I didn't know there was any Jews living on Third Avenue," said the man whose wife didn't live there either.

"Hey," my father shouted after me. "Where are you off to?" (182).

6.   A new look at common ground and relevance

As argued earlier, common ground clearly does get used in many instances of language production. Audience design is something that speakers do. When they don't, many times a meaning negotiation must then take place between the interlocutors, so that conversation and joint activity can continue. Common ground also clearly is used in many instances for comprehension.   Indeed,   if   a   conversational   contribution   is incomprehensible to a hearer because it fails to follow common ground, often the interlocutors will have to repair before continuing.

However, this kind of common ground usage for production and comprehension can also be a problem for interlocutors and conversations. Too tight a reliance on common ground for some production and comprehension can stifle talk and blind understanding. Moreover, if there is an imbalance in the degree of common ground reliance between interlocutors, that can also harm communication. If a hearer, for instance, insists that each and every comment by a speaker is very tightly linked to the current content of common ground, but the speaker is seeking a much more flexible and liberal usage, then the conversation will often go awry. The speaker will feel unduly constrained or the hearer will feel lost or insulted.

Common ground might thus be best viewed as an adjustable component of conversations, both in terms of how much it is needed for a task at hand, as well as how much different speakers wish to rely on it for a given conversational exchange. It does appear more important for some tasks compared to others. Conversations also seem to require a reasonable match in interlocutors' level of reliance on common ground and perhaps also a correspondence in whether that reliance is increased or decreased as the conversation continues. These issues might be addressed by research on how common ground is adjusted for different tasks and by different interlocutors.

The work on memory discussed earlier also may force a rethinking of the centrality of common ground in production and comprehension. An analogy from biology may serve to demonstrate this (see also the use of this analogy in discussing human development in Elman et al. 1998). The arrangement of storage and nursery cells in a honeybee hive was considered for some time a remarkable evolutionary and mathematical achievement on the part of a seemingly simple species. Honeybees seemed to have discovered the perfect and most efficient design for their honeycombs. The maximal amount of both storage space and strength was wrought out of a minimal amount of material and construction labor. It thus seemed that honeybees had invented by decision architecture for their hives that was optimal.

It turns out that honeybees in fact know nothing about optimal design of their hives; rather the honeycomb structure simply emerges from the packing principle, or the accident, if you will, that the structure that emerges by packing circular shapes as tightly as possible just happens to be the hexagonal arrangement in a honeycomb.

Common ground may operate in a similar fashion. Rather than always being a preexisting memory schema that a speaker holds in mind and consults prior to making an utterance, it is a resource that a hearer also consults in the process of determining the meaning of that utterance. Common ground instead could simply be what people come to believe they and their interlocutor must mutually know, after the fact, given that a speaker made a production that was comprehended in a certain way by a hearer.

This view also opens the role of social interaction mechanisms in common ground formation. If common ground is malleable as a form of human memory, then the social relationships among interlocutors can greatly affect what they have in their common ground. A domineering person, for instance, who is greatly admired by many interlocutors, can set the stage for what is discussed and collectively encoded as the common ground. Concomitantly, a person lower in social stature may fail to influence conversation and common ground even if that person has an objectively accurate recall of external events. Anyone who has felt frustrated by the collective discourse, memory, and (mis)representation of recent events, perhaps by a large uninformed populace being led by an incompetent, to put it politely, but versed-in-persuasive-techniques political administration, understands this position very well.

One might also note the potential for cognitive dissonance to play a role in common ground. A person, having expended a lot of effort in a conversation with someone else, may leave that conversation with a delusion that they have developed rich mutual meaning with that other person. Research on cognitive dissonance has repeatedly shown that people seek to align their beliefs and their behavior. If behavior has gone on in such a way that might deviate from beliefs (e.g., a person talked a lot with an interlocutor but no mutual meaning was created), then typically beliefs will be changed to match the behavior (e.g., the interlocutors must have created mutual meaning to justify all the conversational effort).

One last way that common ground might be reconsidered involves work from Robert Bjork and colleagues concerning cognitive effort and retention (Bjork and Bjork, 2006). Very briefly, this work has shown that a short term struggle at comprehending some information will often lead to poor short term measurements of that comprehension, but much better long term retention and comprehension. Short term success at comprehending some information, however, will produce the opposite pattern—high quality short term comprehension measures, but then poor long term retention and comprehension.

Although much of this work has been addressed with educational ramifications in mind, it might also apply to the role common ground plays in language comprehension. If some bit of information does not readily align with the current schematic structure of common ground in an interlocutor, that information is not likely to get integrated into common ground, and is not likely to be thoroughly understood in the short term. A significant time later, however, that information is much more likely to be retained and comprehended. Conversely, information that is readily incorporated into common ground in the short term is less likely to be retained and comprehended over time.

For example, imagine a person is conversing with an interlocutor. The person has currently in his or her common ground representation the memory that the interlocutors had eaten dinner at a particular restaurant with a particular other couple. The person mentions this memory and the interlocutor says one of two different things. The first is an agreement with this memory, followed by an additional mentioning of the fact that there had been a friendly fight over the check (both couples trying to pay for the other). Since the additional fact had been an occasional occurrence in outings with this group of people, the fact is readily assimilated into the person's common ground representation. The other response from the interlocutor is an outright denial that dinner had ever taken place with that couple at that restaurant. This directly conflicts with the person's current common ground representation, and thus is not readily assimilated.

The point of interest for common ground theorizing lies some months later, when the person again considers this portion of their common ground with their interlocutor. At this point, it is more likely that the conflicting information will now be contained in the common ground than the other information that would have been readily assimilated at first encounter. Information that a person struggles to understand will often be more readily understood and retained later, relative to information that was readily understood initially. Common ground may thus over the long term, perhaps somewhat counter-intuitively, be more readily influenced by information that greatly conflicts with it relative to information that confirms it.

7. Conclusion

In its very broadest sense, something akin to common ground must underlie language use and comprehension. Speakers and interlocutors must at least implicitly adhere to some essential coordination guidelines for any degree of shared meaning to happen. Speakers don't normally start talking in different languages that their interlocutors don't know and expect meaning to be shared. Speakers don't normally obliterate turn-taking and other pragmatic rules by talking continuously while their interlocutor is speaking. Speakers don't frequently and regularly use words known to be completely unknown to their interlocutors and expect understanding. Shared meaning doesn't typically happen if interlocutors completely fail to deal with addressees.

On the other hand, the narrower sense of common ground as a sort of strike zone through which speakers *must* produce utterances, and hearers *must* comprehend them, doesn't seem absolutely necessary. Audience design is suspended under many of the talk motivations discussed in this chapter. And comprehension can still proceed even under the most minimal qualifications of optimal relevance (when a hearer realizes an utterance is minimally relevant and interprets it accordingly). True, much of the time common ground *is* used for audience design and relevance assessment; particularly so when interlocutors are diligently negotiating some meaning that is necessary for a joint physical activity (e.g., following directions to assemble some material object). In many other instances speakers don't regularly consider common ground with an interlocutor or audience when they talk; and listeners don't necessarily have to use common ground to gain some comprehension of a speaker's talk. Other mechanisms can fill in, when common ground isn't being fully used, to enable reasonably adequate shared meaning to happen or at least good-enough meaning for the purposes at hand.

Moreover, even in the instances where common ground is being used to marshal and corral shared meaning, it would likely only work as it is currently theoretically described, in reasonably short-term conversations and tasks where memory effects cannot arise to change and erode the content of common ground. If I tell an interlocutor to hand me a certain tool that I label a "pitchfork" for instance, and we then negotiate that I mean the blue-handled three-tined tool and not the red-striped two-tined one, we *can* readily use "pitchfork" in that task to refer to this tool (and "the other one" for the other tool) because we've adequately grounded it. But even in these brief instances, common ground can still be fallible—recall the serial position effects and the example about being available for Tuesday meetings.

Conversely, if there is a wish to treat common ground as a broader construct that explains what I can reliably expect another person to remember that I also remember (and that we both know we remember) over some period of time, then there may a huge problem. Common ground, being a kind of human memory with all the fallibility and malleability well established about human memory, will have to be reconsidered when applied to longer instances of discourse where memory is needed to retain information across contexts. In these longer instances it might be better to consider common ground as a form of cognitive dissonance derived representation—it isn't what I and an interlocutor mutually remember; rather it is what we come to believe we remember given that we've exchanged some meaning. It is still a useful construct for discussing language use and comprehension in these longer time frames, but how it functions there may need reworking.

I've thus advocated that common ground instead be viewed as a variable construct, one whose importance and degree of usage increases when interaction demands it, but can also decrease when other modes of talk and comprehension are taking place. Indeed, there are some modes of talk and interaction where too heavy a reliance on common ground for production and comprehension are stifling.

Related to the common ground as memory view, my students often protest when we discuss the fallibility of human memory, given their rock-solid subjective experiences of their episodic memories. I explain to them that memories that have been somehow altered from an original experience are indistinguishable from memories that are genuinely accurate, to which they still protest, "but how can we ever do anything if our memories are so malleable?" I further explain that what enables functioning is the *illusion* of consistent memories. Experiments can easily show that memory accuracy is not a requirement of daily life—memories are inaccurate yet we somehow still function. The *illusion* of memory accuracy, however, is something we cannot do without. Its necessity indeed explains my students' very protestations. I frequently add the analog of judicial systems—their inaccuracy is easily demonstrated. Innocent people are jailed and guilty people go free. But the *illusion* that judicial systems are fair is indispensible. Were the full mass of people to realize and assimilate that their judicial system isn't fair and accurate, it would soon collapse. Something similar may also be the case with common ground in language. That it is fallible as an objectively anchored memory system is demonstrable. The *illusion* of its full accuracy, and the mechanisms described here that enable that illusion of shared meaning and allow communication to occur anyway, may thus be what sustains common ground.

Note

1. I beg the sophisticated reader's pardon in using these examples from American popular culture. I've jokingly said in conference talks on this topic that if a researcher cannot find examples of their linguistic phenomena in either the Simpsons or Seinfeld, then they don't have their problem figured out (primarily as an off-hand compliment of the creativity of the writing of those programs). So I felt obliged to address my own challenge.

References

Bangerter, Adrian, and Herbert Clark
    2003   Navigating joint projects with dialogue. *Cognitive Science: A Multidisciplinary Journal* 27(2), 195–225.
Bangerter, Adrian, Herbert Clark, and Albert Katz
    2004   Navigating joint projects in telephone conversations. *Discourse Processes* 37(1), 1–23.

Barr, Dale J., and Boaz Keysar
    2005   Making sense of how we make sense: The paradox of egocentrism
                in language use. In *Figurative language comprehension: Social and
                cultural influences*, Herbert Colston and Albert Katz (eds.), 21–41.
                Hillsdale, NJ: Erlbaum.
Bartlett, Fredrick C.
    1932   *Remembering.* Cambridge: Cambridge University Press.
Bjork, Robert, and Elizabeth Bjork
    2006   Optimizing treatment and instruction: Implications of a new theory
                of disuse. *Memory and society: Psychological perspectives.* New
                York: Psychology Press.116–140.
Carreiras, Manuel, Morton A. Gernsbacher, and Victor Villa
    1995   The advantage of first mention in Spanish. *Psychonomic Bulletin &*
                Review 2(1): 124–129.
Carston, Robyn, and George Powell
    2006   *Relevance theory–new directions and developments. The Oxford*
                handbook of philosophy of language. New York: Clarendon
                Press/Oxford University Press. 341–360.
Clark, Herbert, Herbert Schreuder, and Samuel Buttrick
1983   Common ground and the understanding of demonstrative reference.
                Journal of Verbal Learning and Verbal Behavior 22(2): 245–258.
Colston, Herbert L.
    2005   On sociocultural and nonliteral: A synopsis and a prophesy. In
                Figurative language comprehension: Social and cultural influences,
                Herbert L. Colston and Albert .Katz (eds.), 1–20. Hillsdale, NJ:
                Erlbaum.
Colston, Herbert L., and Albert Katz. (eds.).
    2005   *Figurative language comprehension: Social and cultural influences.*
                Hillsdale, NJ: Erlbaum.
Cuc, Alexandru, Yasuhiro Ozuru, David Manier, and William Hirst.
    2006   On the formation of collective memories: The role of a dominant
                narrator. *Memory & Cognition* 34(4): 752–762.
Elman, Jeffrey, Elizabeth Bates, Mark Johnson, Annette Karmiloff-Smith,
                Domenico Parisi, and Kim. Plunkett
    1996   *Rethinking innateness: A connectionist perspective on development.*
                Cambridge, MA: MIT Press.
Ford, Thomas, Charles Stangor, and Changming Duan
    1994   Influence of social category accessibility and category-associated
                trait accessibility on judgments of individuals. *Social Cognition*
                12(2): 149–168.
Gernsbacher, Morton A., and David Hargreaves
    1988   Accessing sentence participants: The advantage of first mention.
                Journal of Memory and Language 27(6): 699–717.

Gernsbacher, Morton A., David Hargreaves, and Mark Beeman
1989   Building and accessing clausal representations: The advantage of
         first mention versus the advantage of clause recency. *Journal of
         Memory and Language* 28(6): 735–755.
Gerrig, Richard J., and William S. Horton
2005   Contextual expressions and common ground. In *Figurative language
         comprehension: Social and cultural influences,* Herbert L. Colston
         and Albert Katz. (eds.), 43–70. Hillsdale, NJ: Erlbaum.
Gibbs, Raymond, and Markus Tendahl
2006   Cognitive effort and effects in metaphor comprehension: Relevance
         theory and psycholinguistics. *Mind & Language* 21(3): 379–403.
Giora, Rachel
2003   *On our mind: Salience, context, and figurative language.* New York:
         Oxford University Press.
Gregory, Michelle L., Alice F. Healy, and Daniel Jurafsky
2002   Common Ground in Production: Effects of Mutual Knowledge on
         Word Duration. Unpublished Manuscript. University of Colorado at
         Boulder.
Herr, Paul
1986   Consequences of priming: Judgment and behavior. *Journal of
         Personality and Social Psychology,* 51(6): 1106–1115.
Horton, William, and Richard Gerrig
2005   Conversational common ground and memory processes in language
         production. *Discourse Processes* 40(1): 1–35.
Jarvikivi, Juhani, Roger van Gompel, Jukka Hyona, and Raymond Bertram
2005   Ambiguous pronoun resolution: Contrasting the first-mention and
         subject-preference accounts. *Psychological Science* 16(4): 260–264.
Jaszczolt, Katarzyna
2005   *Default Semantics.* Oxford: Oxford University Press.
Kecskes, Istvan
2007   Formulaic language in English Lingua Franca. In *Explorations in
         Pragmatics: Linguistic, Cognitive and Intercultural Aspects,* Istvan
         Kecskes, and Laurence Horn (eds.), 191–219. Berlin/New York:
         Mouton de Gruyter:
Keysar, Boaz, and Bridget Bly
1995   Intuitions of the transparency of idioms: Can one keep a secret by
         spilling the beans? *Journal of Memory and Language* 34(1): 89–109.
Kim, Sung-il, Jae-ho Lee, and Morton A. Gernsbacher
2004   The advantage of first mention in Korean: The temporal
         contributions of syntactic, semantic, and pragmatic factors. *Journal
         of Psycholinguistic Research* 33(6): 475–491.

Kogen, Jay, and Wallace Wolodarsky
    1993   Last exit to Springfield. *The Simpsons,* Matt Groenig (Producer).
                Los Angeles: Fox Broadcasting Company.
Loftus, Elizabeth, and Larry Cahill
    2007   Memory distortion: From misinformation to rich false memory. In
                *The foundations of remembering: Essays in honor of Henry L.
                Roediger III,* James S. Nairne (ed.), 413–425. New York:
                Psychology Press.
Long, Tim, George Meyer, Mike Scully, and Matt Selma
    2000   Behind the laughter. *The Simpsons,* Matt Groenig (Producer). Los
                Angeles: Fox Broadcasting Company.
Mandel, David, and Peter Mehlman
    1997   The betrayal. *Seinfeld.* Larry David and Jerry Seinfeld (Producers).
                Los Angeles: National Broadcasting Company.
McDonald, Janet, and Deborah Shaibe
    2002   The accessibility of characters in single sentences: proper names,
                common nouns, and first mention. *Psychonomic Bulletin & Review*
                9(2): 356–361.
Mey, Jacob
    2001   *Pragmatics: An introduction.* Oxford: Blackwell.
Murdoch, Bennet
    1962   The serial position effect of free recall. *Journal of Experimental
                Psychology* 64: 482–488.
Oakley, Bill and Josh Weinstein
    1994   Lady Bouvier's lover. *The Simpsons,* Matt Groenig (Producer). Los
                Angeles: Fox Broadcasting Company.
Russo, Richard
    1988   *The risk pool.* New York: Random House.
Smith, Eliot, and Nyla Branscombe
    1988   Category accessibility as implicit memory. *Journal of Experimental
                Social Psychology* 24(6): 490–504.
Smith, Sara, Hiromi Noda, Steven Andrews, and Andreas Jucker
    2005   Setting the stage: How speakers prepare listeners for the introduction
                of referents in dialogues and monologues. *Journal of Pragmatics*
                37(11): 1865–1895.
Sperber, Daniel
    2005   Modularity and relevance: How can a massively modular mind be
                flexible and context-sensitive? In *The innate mind: Structure and
                contents,* Peter Carruthers, Stephen Laurence, and Stephen Stich
                (eds.), 53–68. New York: Oxford University Press.
Stalnaker, Robert
    1978   Assertion, In *Syntax and Semantics* 9: *Pragmatics,* Peter Cole (ed.),
                315–322. New York: Academic Press

Weldon, Mary S.
2001 Remembering as a social process. In *The psychology of learning and motivation: Advances in research and theory, Vol. 40*, Douglas L. Medin (ed.), 67–120. San Diego: Academic Press.

A memory-based approach to common ground and audience design

William S. Horton

1. Introduction

Individuals in dialogue frequently make reference to a variety of topics during the course of typical interactions. Most of the time, the referring expressions that speakers produce work well, in the sense that addressees give little indication of any difficulties identifying the intended referents. This general pattern of success might give the impression that reference is an easily managed aspect of discourse processing. Yet, as with other aspects of language production, instances in which speakers produce errors are quite useful for revealing the underlying complexities of successful reference generation. Consider, for example, the following moments taken from the Call Home Corpus, a collection of telephone conversations between friends and family members collected by the Linguistic Data Consortium (Kingsbury et al. 1997):

(1) A: And one of her students showed her how to get into *the X-500 directories*.
B: Which are?
A: Hm?
B: What are the X-500 directories?
A: Oh um where you put- your um- How c- How can you not know?

(2) B: Last weekend Lida and Irv Teisher were in town.
A: Oh.
B: They uh- they have a son who is the uh- I guess he buys books for *something called Borders*, which is a bookstore that-
A: Yes, we have it here too.

Both of these conversational excerpts contain infelicitous referring expressions. In example (1), speaker A appears to mistakenly attribute too much knowledge to her addressee, referring to the "X-500 directories" without further identifying information, which prompts the addressee to seek clarification about what is meant. In example (2), however, speaker B appears to attribute too little knowledge to her addressee, introducing the referent as "something called Borders" until the addressee interrupts to correct the assumption that Borders is unfamiliar. Moments such as these are useful because they make explicit the fact that most conversations play out against a vast background of assumptions concerning the knowledge of others. Much of the time, these assumptions go unnoticed because, for the most part, conversations proceed remarkably smoothly–people refer to numerous entities and concepts without causing overt problems for their interlocutors. It is primarily when addressees find it necessary to reply, "What do you mean?" or "Yes, I know that already," that the challenges faced by speakers in formulating referring expressions become more apparent. If a speaker is overly specific, she risks being insulting; if a speaker is too vague, she risks leaving addressees confused. Being a cooperative conversationalist, then, must involve, at some level, taking into consideration the information available to one's interlocutor (Grice 1975).

To explain how interlocutors successfully navigate these complexities in conversation, theories of language use have typically appealed to the notion of *common ground*, which refers to the set of knowledge and beliefs taken as shared between interlocutors. Individuals are assumed to interact on the basis of beliefs about their common ground, which inform decisions about when and how to refer to objects and entities for other people, and how to understand such references (Clark 1996). In the domain of language production, the manner in which speakers tailor utterances to reflect considerations of shared knowledge is known as *audience design* (Clark and Murphy 1982; for reviews, see Schober and Brennan 2003; Barr and Keysar 2006). Salient evidence for audience-driven modifications to speech comes from situations in which individuals make one-off adjustments to global characteristics of their utterances, such as overall complexity, in response to the perceived needs of particular classes of addresses (e.g., "Motherese," or child-directed speech [Snow and Ferguson 1977] and "elderspeak," or speech directed toward the elderly [Kemper 1994]). Even more frequently, however, cooperative speakers are called upon to make relatively fine-grained adjustments to features of their utterances based on knowledge about particular individuals (Dell and Brown 1991). Under these circumstances, according to the tenets of audience design, speakers should avoid terms like "X-500 directories" unless they have reason to believe that the requisite information is also available to their addressees.

In general, then, whether people show evidence of audience design appears to depend greatly on access to beliefs about common ground. In this chapter, I will briefly describe some of the primary attempts within psycholinguistics to explain how language users manage the complexities of considering information relevant to conversational common ground. Then, I will outline a recent account proposed by myself and Richard Gerrig in which we suggest that additional traction on the problem of common ground can be gained by considering the possible role played by ordinary cognitive mechanisms of memory encoding and retrieval (Horton and Gerrig 2005a). One implication of this memory-based view will be that conversational phenomena like audience design may not necessarily involve explicit computations of shared knowledge. Then, I will summarize some recent work carried out under the auspices of this memory-based approach. This research has demonstrated how conversational coordination and audience design in particular are influenced by how communicative situations facilitate memory encoding and retrieval, and also by the memory capacities of particular speakers. In general, the aim of this program of research has been to "ground" common ground by exploring the extent to which it can be mediated through more fundamental cognitive psychological mechanisms.

2.   Previous accounts of common ground inference in conversation

Although common ground is a central concept for language use, it is important to appreciate that *actual* (i.e., veridical) knowledge of what information is in common ground is in some sense impossible, given that we never have direct access to the knowledge possessed by others. Moreover, because the concept of common ground includes knowledge about another's knowledge, by definition it must also include information about *mutual knowledge*, which refers to the set of knowledge that is mutually known to be shared by all parties in an interaction (Schiffer 1972). For example, both speakers might independently have knowledge of the X-500 directories, but in order for a speaker to felicitously refer to the "X-500 directories," she would have to not only believe that her addressee knows this information (i.e., that it is part of their common ground), but would also have to believe that her addressee knows that she believes this. Even this, however, would not be enough to guarantee mutual knowledge, because her addressee would also have to believe that the speaker believes that the addressee knows about this belief—and so forth. As Schiffer (1972) demonstrated, this chain of reasoning necessary for "true" mutual knowledge quickly leads to an infinite regress, termed the *mutual knowledge paradox*.

From the viewpoint of trying to understand the cognitive processes that enable language use, the concept of mutual knowledge clearly poses a challenge. Successful interactions appear to require mutual knowledge, yet mutual knowledge represents a psychologically implausible state of affairs. There have been a number of attempts to resolve this paradox, both in philosophy and linguistics (see Smith 1981 and Lee 1999 for discussions). Within psycholinguistics, the most influential attempt at a solution to this problem was proposed by Clark and Marshall (1981) and further developed through the work of Clark and colleagues (Clark 1992). Clark and Marshall's account drew upon the insights of Lewis (1969), who argued that people derive beliefs about shared knowledge to the extent that there are appropriate *bases* for doing so. Lewis proposed that such bases include things like prior agreements (i.e., conventions) and contextual salience, and that these aspects of interpersonal contexts provide individuals with reasons to assume that particular types of knowledge might be shared with other individuals in those same contexts. For example, if it is conventional within a particular community for a nod of the head up and down to mean "yes," then knowing that someone else is a member of the same community will provide a useful heuristic for assuming that this convention is shared knowledge with that person, even when this fact has not been independently established.

Following the arguments put forth by Lewis (1969) and Schiffer (1972), Clark and Marshall (1981) proposed that interlocutors rely upon a set of heuristics for assuming that particular knowledge can be taken as shared. Their particular insight lay in proposing that these rules of thumb rely upon considerations of "co-presence" between speakers and addressees. Specifically, interlocutors are assumed to truncate the infinite regress implied by the mutual knowledge paradox by seeking evidence for *triple co-presence*, in which the trio of speaker, addressee, and referent are all "openly present together" (Clark and Marshall 1981: 32). Clark and Marshall described triple co-presence as applying in three domains. First, *physical co-presence* refers to information that is in the shared physical or perceptual environment of the interlocutors. Next, *linguistic co-presence* refers to information that can be derived from past and present conversations between interlocutors. Finally, *community membership* refers to information that is part of the interlocutors' shared socio-cultural background. Importantly, each of these co-presence heuristics provide a basis for common ground inference, through which speakers and addressees assume that information that meets one or more of the requirements for co-presence can in fact be treated as mutually known.

In their original discussion, Clark and Marshall (1981) argued that the complexities of definite reference demand that people possess a special type of memory representation that directly encodes whether particular events meet the standard of triple co-presence. They called this type of representation a *reference diary* (Clark and Marshall 1978), defined as "a log of those events we have personally experienced or taken part in with others" (Clark 1996: 114). These special person-centered discourse representations were intended to capture aspects of mutually experienced situations relevant for evaluating co-presence. For example, your reference diary for your friend Mary might contain a record of the fact that she was also present at the holiday party when Bob knocked over the punch bowl. If you wished to discuss this event with Mary subsequently, you would consult your relevant diary entry to discover strong evidence that she also knows about Bob's mishap, which would in turn shape how you referred to this event with her. According to implications of this *Optimal Design* view (Clark, Schreuder, and Buttrick 1983), speakers are expected to routinely tailor their utterances for particular addressees by searching reference diaries for evidence to assure themselves of triple co-presence. Similarly, listeners routinely assume that speakers' utterances were optimally designed with their needs in mind.

Clark and Marshall's (1981) proposals represent an appealing solution to the problem of mutual knowledge on several fronts. In particular, the co-presence heuristics capture something important about the types of information that might regularly support inferences about the knowledge of others. For example, once interlocutors have successfully established a meaning for an expression like "X-500 directories" (a process that Clark and colleagues have referred to as "grounding"; Clark 1996; Clark and Wilkes-Gibbs 1986), this meaning can be taken as linguistically co-present for the purpose of further interactions. Additionally, the notion of reference diaries highlights the fact that memory encoding and memory retrieval have a critical role to play in any description of conversational common ground. Knowledge related to one's experiences with and beliefs about other individuals clearly must be stored and retrieved from memory in some fashion – there are no other reasonable alternatives.

There are at least two reasons, however, why the particular solution provided by reference diaries is unsatisfactory. On the representational side, Optimal Design entails the strong assumption that individuals maintain highly detailed records of their interlocutors, and that these representations are always available to shape language use. It is not clear, however, how one would know *a priori* what the "right" level of detail should be such that adequate evidence of triple co-presence would be available under most circumstances. If reference diaries encoded triple co-presence in every instance possible, they would quickly become representationally unbounded. If they encoded information selectively, then it isn't obvious what selection criteria would apply. On the processing side, the Optimal Design view doesn't attempt to provide an explanation for *how* beliefs about common ground, inferred on the basis of evidence encoded in reference diaries, become incorporated into language use. The primary focus of Clark and Marshall's (1981) proposal was in providing a description of the kinds of information relevant for common ground inference in instances of definite reference. From a cognitive psychological standpoint, though, any adequate model of conversational common ground must describe not only the memory representations that potentially encode information relevant for common ground inference, but also the cognitive mechanisms responsible for constraining the types of information considered by language users as they interact with others.

Partially in response to these issues, an alternative perspective on the role played by common ground in conversation has been developed by Keysar and colleagues (Keysar et al. 1998; Keysar et al. 2000). This *Perspective Adjustment* account explicitly rejects the claim, inherent in the Optimal Design view, that information relevant to common ground is necessarily taken into account from the earliest moments of language processing. Instead, it is assumed that initial aspects are carried out without explicit consideration of one's conversational partner. This position is motivated by the supposition that incorporating beliefs about the knowledge and perspectives of others is an effortful process that requires additional time and cognitive resources beyond what is necessary for routine aspects of utterance planning and interpretation.

An important implication of the Perspective Adjustment view is that language use has an egocentric basis, given that one's *own* knowledge will typically form the starting point for any given instance of utterance planning or interpretation (Keysar, Barr, and Horton 1998). Advantageously, this provides a highly testable description of the time course with which one should expect information relevant to common ground to exert an impact upon language processing, and in a number of studies Keysar and colleagues have obtained evidence consistent with an egocentric bias during the earliest moments of processing. For example, Keysar et al. (2000) used eye-tracking methodology to show that addressees were likely to initially rely on their own perspectives when considering which object in a display was the intended referent of an expression produced by their partner. When the display contained a plausible referent that was visually available *only* to the addressee, analyses of the addressees' eye movements revealed that the presence of this privileged object momentarily interfered with successful identification of the intended referent, which was always mutually visible. These results argue against the strong claim, implied by the Optimal Design account, that beliefs about common ground necessarily act as an immediate constraint upon language use (Barr and Keysar 2006).

Horton and Keysar (1996) reported analogous evidence for perspective adjustment in the domain of language production. In that study, speakers described visual objects (e.g., a medium-sized triangle) for addressees who either did or did not share critical context information (e.g., a smaller triangle). Optimal design would predict that speakers should refer to a target object with a context-relevant modifier (e.g., big triangle) only when the context information is simultaneously available to the addressee. A referring expression like "big triangle" has little meaning except when the relevant contrast set is part of the information shared between speaker and addressee (Olson 1970). With no constraints upon the time available for utterance planning, this is exactly what speakers did – they produced more descriptions like "big triangle" when the context was shared than when it was privileged. When placed under pressure to produce descriptions quickly, however, speakers failed to show the same sensitivity to the addressees' knowledge – the levels of context-relevant modification were similar regardless of whether the context information was shared or not. These results were interpreted as suggesting that taking into account the information available to one's addressee is a slower, more effortful process that is readily disrupted by factors such as cognitive load.

For current purposes, the important aspect of the process model described by the Perspective Adjustment approach is that the mechanisms responsible for taking into account the knowledge and perspectives of others are seen as acting separately from other facets of routine language processing. As expressed by Barr and Keysar (2006), "common ground is a functionally distinct process that belongs to an 'adjustment' stage of processing, but that it imposes no constraint on production or comprehension processes per se" (904). On this view, common ground inference is optional in the sense that it is primarily directed at error detection and correction. In this respect, it shares with the Optimal Design view an emphasis upon the specialized nature of the cognitive mechanisms responsible for accommodating to beliefs about common ground. Whereas Clark and Marshall's (1981) notion of "reference diaries" presumed the existence of special-purpose memory representations, the Perspective Adjustment view presumes that adjusting to the perspectives of others occurs via a dedicated set of processes. One goal of the memory-based view outlined in this chapter is to obviate the need to appeal to special processes or special representations to explain every circumstance in which common ground inference is evident.

3.   Memory-based processing and audience design

In contrast to these previous accounts, then, Richard Gerrig and I recently proposed an account of common ground inference in language production that emphasizes the role played by ordinary processes of memory encoding and retrieval (Horton and Gerrig 2005a).   We began by drawing a distinction between two different aspects of audience design that have distinct roles for understanding how speakers come to produce utterances appropriate for particular addressees: *commonality assessment* and *message formation*. We intended commonality assessment to refer to the means by which beliefs about common ground are made manifest to language users, while message formation refers more specifically to the processes by which speakers shape utterances to suit addressees (an analogous process in language comprehension might be termed *message interpretation*).   To better understand the utility of this distinction, imagine one of two possible responses to an expression like "I saw *Percival* the other day."   An inadequate assessment of commonality on the part of the speaker might elicit the response, "Who is Percival?" (i.e., Percival is not in common ground), whereas inadequate message formation might elicit "Which Percival?" (i.e., it isn't clear which of several Percivals is meant). Note that in the latter circumstance, there is no confusion about the status of the intended referent with respect to common ground; the issue is simply how this information comes to be reflected in utterance planning.   In short, commonality assessment describes how language users derive beliefs about common ground while message formation describes how these beliefs influence utterances.

Most of the psycholinguistic research examining conversational common ground has focused on aspects of message interpretation and message formation, rather than commonality assessment per se. Specifically, researchers in various ways have obviated the need for separate assessments of commonality through experimental manipulations that specify in the discourse context whether particular information can be taken as "shared" or "privileged." As described previously, work in this area has typically been concerned with whether speakers and addressees take these divisions into account (e.g., Barr and Keysar 2002; Brennan and Clark 1996; Hanna, Tanenhaus, and Trueswell 2003; Horton and Gerrig 2002, 2005b; Horton and Keysar 1996; Keysar et al. 2000; Nadig and Sedivy 2002). For example, in Brennan and Clark (1996), speakers negotiated suitable referring expressions for target objects appearing in the context of other objects of the same kind (e.g., referring to "the blue car" in the presence of multiple cars). When the context was changed to make the target objects unique, speakers reverted to simple, basic-level expressions (e.g., "car"), but did so more quickly when the partner changed compared to when the partner stayed the same. In these circumstances, the previously established referring expressions were clearly part of the speakers' privileged knowledge when the partner was new, but were shared knowledge when the partner was old (in Brennan and Clark's terms, they were part of the "conceptual pact" established between partners). The salience of this partner change presumably influenced the speakers' decisions (whether implicitly or explicitly) about how to formulate referring expressions to suit specific partners' communicative needs.

The notion of commonality assessment, in contrast, refers to processes that describe how partner-related information becomes accessible to language users in the first place. In Horton and Gerrig (2005a), we argued that commonality assessment is likely to occur via both automatic and strategic mechanisms. With respect to strategic commonality assessment, one can find many conversational situations in which interlocutors appear to explicitly engage in an effortful search of memory for evidence concerning common ground. For example, consider this excerpt from the Call Home corpus:

(3) Yeah, I've got another buddy who, uh, is a Marine pilot. I'm trying to think if you had ever met this guy. I don't think so.

In such moments, interlocutors appear to strategically assess the likelihood that particular information is shared. These controlled assessments may be triggered by particular feedback from one's conversational partner, by situations that require careful attention to distinctions between shared and privileged knowledge (e.g., surprise parties or secrets), or by specific interpersonal motivations such as politeness goals and sensitivity to social status. In these circumstances, language users may rely upon memory representations that potentially function very much like Clark and Marshall's (1981) notion of a reference diary, in the sense that they provide evidence about the status of particular information *vis a vis* common ground. Similarly, these moments presumably also frequently involve explicit adjustments to the perspective of one's partner based on the results of strategic considerations of common ground. In general, though, the memory representations relevant for strategic commonality assessment are more likely constructed on the spot instead of simply retrieved from memory.

While the absolute importance of strategic assessments of common ground remains very much an open issue, the rapidity and the fluidity of most natural conversations suggest that strategic commonality assessment may represent the exception rather than the norm. As we argued in Horton and Gerrig (2005a), it is likely to be the case that commonality assessment, as a process of deriving beliefs about whether particular information can be taken as part of common ground, functions more routinely in an automatic fashion. Cognitive psychological research on memory, and episodic memory in particular (Tulving 2002), provides a starting point for understanding how automatic commonality assessment might work. Episodic traces, as memories of experienced events that occur in a specific time and place, potentially capture a variety of information about people's interactions with others. Importantly, these memory representations do not have to be goal-driven in the sense of being constructed to provide particular evidence for co-presence. Rather, they simply serve as records of the contextual details that permeate life experiences. It is this routine episodic encoding that enables other individuals, who are a part of those experiences, to become linked, or associated in memory, to a wide range of related information. Once these connections have been established, those same individuals, when encountered subsequently, can then serve as salient memory cues for the automatic retrieval of entire patterns of associated information.

This memory retrieval is conceptualized as taking place through a process known as *resonance*, which is a cue-based search that occurs in parallel throughout the contents of long-term memory. Resonance is a fast, passive, and effort-free mechanism in which cues in working memory interact in parallel with information residing in long-term memory (Ratcliff 1978). Because resonance provides a parallel search of memory, it is possible for a wide range of associated information to become accessible on the basis of relatively local cues. Importantly, resonance has been implicated as a process that functions quite broadly in a variety of situations. For example, memory-based accounts of narrative comprehension have described how resonance on the basis of cues presented in texts can cause associated information to become more immediately accessible even when this information has been fully backgrounded by intervening material (e.g., Albrecht and Myers, 1998; Gerrig and McKoon 1998; Lea et al. 1998).

The application of resonance to conversational contexts is inspired in part by existing global-matching models of recognition memory (e.g., Gillund and Schiffrin 1984; Hintzman 1986; Ratcliff 1978). These models provide a possible description of how information associated with particular conversational partners could become more immediately accessible in memory. For example, SAM (Search of Associative Memory; Gillund and Schiffrin 1984) and its more recent incarnation REM (Schiffrin and Steyvers 1997) conceptualize memory retrieval as a cue-dependent global search of long-term memory. According to REM, contextual information available at the time of encoding is highly likely to become part of relevant memory traces. If this context information matches cues present at the time of search, then the likelihood of successful retrieval will increase. In these models, then, increased accessibility of context-relevant information can occur simply as the result of overlap between associations present during encoding and retrieval (Pecher and Raaijmakers 2004).

Evidence for this kind of implicit influence of contextual information on retrieval comes from cognitive psychological research across a variety of domains. For example, episodic priming paradigms have shown that participants are faster to name a target word when it is repeatedly preceded by the same prime than when it is preceded by a different prime upon each presentation (e.g., McKoon and Ratcliff 1979; Spieler and Balota 1996). This repetition of prime-target pairings established implicit contextual encodings that facilitated word recognition. Contextual facilitation can also extend to visual contexts. Using a visual search paradigm, Chun and Jiang

(2003) found that individuals who searched for visual targets in the context of specific configurations of distracter objects were better on subsequent trials at finding new targets embedded in the same contextual configurations, compared to targets in new configurations. These repeated configurations appeared to act as a contextual cue for the associated target locations. Importantly, participants showed no evidence of any conscious recollection for specific configurations.

Similar influence of contextual associations on retrieval can be also found in work on encoding specificity and context-dependent memory, which has shown that information learned under particular conditions is recalled better if those same conditions are reinstated at the time of retrieval (Smith 1994; Tulving 1983). Although context-dependent memory has been observed most frequently in tasks that measure explicit recall, reinstatement of the same context at testing has been shown to facilitate conceptual implicit memory as well (Parker, Gellatly, and Waterman 1999). In the memory-based view of common ground, then, conversational partners potentially act as contextual cues for the automatic retrieval of associated information just as different rooms or different physical contexts can facilitate memory depending on the type of overlap with the context of encoding (Horton 2007). This associative overlap provides an important basis for resonance processes to increase the accessibility of partner-related information in ways that can act as a constraint upon concurrent language processing.

The central implication of this memory based account is that if one has a strong-enough pattern of associations between a conversational partner and some set of relevant information, then the likelihood is high that this information will be taken as being shared knowledge. In this manner, automatic commonality assessment provides one possible basis upon which language users may generate inferences about common ground. Significantly, this is a much weaker standard than having information about triple co-presence encoded directly into a reference diary-like memory representation (Horton and Gerrig 2005a). Obviously, the mere presence of an association will seldom be enough by itself to ensure common ground, and there are many circumstances under which more strategic assessments of commonality may be necessary. Even so, the demands of fluent conversation are likely to encourage language users to process utterances on the basis of whatever partner-relevant information is most accessible at each moment in time.

An advantage of this approach is that it combines particular features of existing accounts of conversational common ground. Like Optimal Design, the memory-based view states that, given the availability of suitable contextual cues and partner-specific memory associations of sufficient strength, beliefs about common ground may indeed serve as an immediate constraint upon processing. But like the Perspective Adjustment view, language use can also be seen as opportunistic, relying upon whatever information is most immediately available, particularly one's own knowledge. Rather than requiring dedicated representations or special processes to account for all conversational uses of common ground, though, the memory-based account assumes that domain-general processes of memory encoding and retrieval provide one important route through which information related to other individuals can influence language processing. As such, this perspective allows common ground to be integrated within a wider range of related cognitive psychological phenomena.

This memory-based solution to the problem posed by common ground shares important features with the recent *interactive alignment* account of conversation proposed by Pickering and Garrod (2004, 2006). Pickering and Garrod suggest that conversational coordination is made possible by low-level mechanisms of alignment that function to enhance the similarity of mental representations across partners in dialogue. On this account, automatic priming processes increase the likelihood that particular behaviors, once produced by an individual in conversation, will be produced more frequently not only by that same individual again, but also by a conversational partner as well. For example, syntactic priming paradigms have shown that individuals are more likely to use a particular syntactic construction like the passive if they have recently been exposed to that construction through their own productions (Bock 1986) or through comprehending the utterances of others in dialogue (Branigan, Pickering, and Cleland 2000). Through these automatic priming mechanisms, which function simultaneously across multiple domains (e.g., syntactic, conceptual, lexical, phonological), the cognitive representations of interlocutors are thought to become "aligned" over the course of an interaction, enabling conversational coordination without the need for any sort of detailed record of each other's knowledge. Interactive alignment shares with the memory-based account proposed by Horton and Gerrig (2005a) the view that many aspects of conversational interactions can emerge on the basis of relatively low-level processes. Whereas interactive alignment emphasizes the role of alignment of representations currently being primed by one's conversational partner, the memory-based approach emphasizes the role of contextual cues and resonance in the automatic retrieval of information from long-term memory.

4. Demonstrating the automatic impact of partner-specific memory associations

In the remainder of this chapter I will focus on several recent studies that support different aspects of this memory-based approach to common ground. As mentioned previously, research into the processes underlying audience design has typically been concerned with issues surrounding message formation, or how speakers formulate utterances in ways that reflect beliefs about common ground. In general, despite the fundamental importance of understanding how beliefs relevant for common ground inference become accessible to speakers in conversation, less attention has been paid to aspects of language processing more directly relevant to commonality assessment proper. The memory-based perspective, though, provides a description of commonality assessment that is rooted in domain-general encoding and retrieval mechanisms that apply quite broadly, even outside of conversational contexts. Because resonance is a cue-driven memory process that is, by definition, not intrinsically goal-directed (Gerrig and O'Brien 2005), specific partner-related memory associations should exert an automatic influence on language production even in the absence of an explicit intent to communicate—as long as the presence of other individuals is sufficient to increase the accessibility of associated memory traces.

To provide evidence to support this conceptualization of automatic commonality assessment, I carried out a study (Horton 2007) that borrowed the logic of standard implicit memory paradigms. Specifically, I examined whether performance on a basic, language production task – picture naming – would be facilitated by the presence of specific individuals associated with the objects being named. Both experiments began with an initial task phase designed to foster the creation of memory associations that would be specific to particular experimental "partners." In this task, participants generated category exemplars based on cues provided by two different individuals, each of whom was present only for one block of trials. The correct answer on each trial was constrained by a partial word fragment on a computer screen. For example, after receiving the category cue "*a musical instrument,"* if participants saw "B_ _J O" on the computer screen they would be expected to produce the response *banjo.* The fact that participants responded to different items with each partner was crucial, because this afforded them the opportunity to encode distinct associations in memory with respect to each partner context.

The second task phase assessed whether these partner-specific memory associations would exert an influence upon language production. In this phase, participants named a series of images of familiar objects presented via computer in the context of each of the same two partners. The partners again were present for separate blocks of trials, and the experimental trials in each block all involved objects that belonged to categories named during the first task phase (e.g., a picture of a banjo). Critically, participants named half of the experimental items in the context of the partner previously associated with that object, while they named the remaining experimental items in the context of the other partner. Despite the fact that this was explicitly not a communicative task – the partners didn't do anything except control the presentation of the pictures – participants were faster to name pictures associated with the individual currently serving as the experimental partner compared to pictures associated with the other partner. This occurred both when these associations existed between specific partners and object labels (e.g., "harp" vs. "banjo;" Experiment 1) and when they existed between partners and entire object categories (e.g., "musical instruments" vs. "birds;" Experiment 2). Furthermore, the strength of this partner-specific priming effect across participants did not correlate significantly with their explicit recall of partner-item associations in a final source memory task. These results demonstrate how domain-general memory processes like resonance can serve to increase the accessibility of information associated with particular interpersonal contexts in ways that impact concurrent language processing. Such automatic "assessments" of commonality are potentially an important means (but presumably not the only means) by which message planning and interpretation are shaped by specific interpersonal contexts.

This demonstration of the influence of partner-specific memory associations on language production is consistent with recent constraint-based accounts of common ground processing (Hanna et al. 2003; Nadig and Sedivy 2002). These accounts, rooted in more general constraint-based models of language processing (e.g., MacDonald, Pearlmutter, and Seidenberg 1994; McRae, Spivey-Knowlton, and Tanenhaus 1998), presume that partner-relevant information is one of many cues that are integrated simultaneously during language interpretation. On this view, common ground knowledge acts as a probabilistic constraint that depends on factors such as contextual salience. Evidence for this view comes from work by Hanna et al. (2003), who monitored the eye movements of individuals as they carried out spoken instructions to manipulate physical objects. Hanna et al. found that, while there was an immediate preference to interpret the instructions as referring to objects in common ground, addressees still experienced momentary interference from information in privileged ground. Thus, it appeared that common ground knowledge only served as a partial constraint upon interpretation due to the competing cues present in the situation. In a similar fashion, information associated with one's conversational partner, retrieved from memory on the basis of low-level processes like resonance, may serve as another probabilistic influence on message planning and formulation (Horton and Gerrig 2005a).

5. Situation-specific variation in memory-based processes of audience design

Similar to the constraint-based view supported by the results of Hanna et al. (2003), the memory-based approach to common ground predicts that people will show evidence for sensitivity to common ground to the extent that particular situations provide appropriate cues to support partner-specific encoding and retrieval. This is in contrast to the all-or-nothing perspective implied by the comparison between the Optimal Design and Perspective Adjustment accounts of common ground inference. To demonstrate how evidence for audience design can depend on the nature of the experiences given to speakers during the course of their interactions with others, Horton and Gerrig (2002) used a variant on standard referential communication paradigms (Clark and Wilkes-Gibbs 1986; Krauss and Weinheimer 1964, 1966) in which triads of participants were asked to carry out a card-matching task that required one person to play the role of "Director" and the other two individuals to be "Matchers." The Director's task was to repeatedly help the Matchers place sets of cards into particular goal arrangements, and participants were prevented from seeing one another's cards by means of visual barriers. Typically, this type of repeated collaboration task produces a set of distinctive conversational behaviors. As partners work together to negotiate particular perspectives on the task materials, these perspectives become part of their established common ground. Once this information has been grounded, the interlocutors can then make use of these shared perspectives on subsequent trials of the task, resulting in shorter, more consistent descriptions over time, with briefer exchanges required to identify each item (Clark and Wilkes-Gibbs 1986)

In Horton and Gerrig (2002), we used two types of cards as materials. Half of the cards depicted abstract shapes known as Tangrams, while the other cards contained full-color photographs of living things. Both types of cards were further subdivided into subcategories: *birds*, *fish*, and *flowers* for the living things and figures that looked like *boats*, *rockets*, and *people* for the Tangrams. Because the Tangrams constituted an unfamiliar referential domain, they were expected to elicit relatively more discussion between Directors and Matchers during the early rounds of the task compared to the living things, which were more familiar and therefore readily describable.

Because we were interested in the nature of the conversational experiences that would influence the tendency for speakers to engage in audience design, the initial phase of the study varied the experiences of participants. In the early rounds of the task, Directors described the complete set of items for both Matchers simultaneously. We distributed the card categories across Matchers, however, such that each Matcher possessed incomplete and partially overlapping sets of cards. Two of the card subcategories (e.g., the fish and rockets) were given exclusively to one Matcher while two other card subcategories (e.g., the flowers and people) were given exclusively to the other Matcher. The remaining two subcategories, though, were given to both Matchers (e.g., the birds and boats). During the task, we instructed the Matchers to listen to the Director's descriptions and to arrange only those cards that were relevant to them, ignoring the cards that they didn't have. In this manner, Directors negotiated suitable referring expressions for particular cards with only *one* Matcher or with both Matchers simultaneously. In later rounds, the Directors carried out the same matching task with each Matcher separately, and in these rounds we gave each Matcher the complete set of cards, only some of which they had seen before. We were interested in whether the Directors' referring expressions would reflect their prior experiences of having previously described certain subcategories of cards with particular Matchers. To examine evidence for audience design, we measured the amount of information Directors chose to produce in their descriptions, the frequency with which their descriptions expressed uncertainty through hedges such as "kinda" or "maybe," and how often they modified their previous descriptions for the same referents by adding new information.

In general, we found that, when describing items that the current Matcher had *not* seen previously (compared to those that the Matcher *had* seen), speakers produced longer descriptions, more hedges, and more modifications of earlier descriptions, consistent with the predictions of audience design.   Importantly, however, these effects were more pronounced for Tangrams and also for the descriptions for the final Matcher. We interpreted this pattern of results as reflecting the nature of the Directors' experiences during the task. Because the Tangrams had elicited the most negotiation initially, Directors were better prepared to demonstrate audience design when describing these items for specific addressees.   From the viewpoint of memory encoding and retrieval, providing speakers with appropriate opportunities to encode information with respect to particular partners allowed them to make use of this information subsequently to design utterances to reflect those experiences. The initial difficulty of describing the Tangrams appeared to have had the ultimate benefit of providing Directors with more specific information about individual Matcher's perspectives, whereas the relatively straightforward descriptions available for the pictures of living things afforded less opportunity for Matcher-specific encoding.   Additionally, Directors' experiences of describing cards that were unfamiliar for the first Matcher seemed to have allowed them to more readily adjust their descriptions appropriately with the second Matcher, presumably because feedback from the first Matcher prompted more careful consideration of addressee-specific information.   These results demonstrate how particular conversational contexts may prompt more strategic consideration of the need for audience design.

In Horton and Gerrig (2002), we varied the specific subcategories of cards that were available to each Matcher during the initial rounds of the task.   We anticipated that this structure would enhance the capacity of Directors to encode partner-specific information.   As we found out, however, participants often failed to perceive the groupings that we used to create the sets of Tangrams in our study, choosing completely unrelated perspectives instead.   We speculated that this lack of a clear category-partner correspondence may have increased the difficulty of encoding suitable partner-specific perspectives for these items, forcing Directors to evaluate the appropriateness of particular conceptualizations on something approaching an item-by-item basis.   To the extent that Directors were less able to rely upon suitable partner-specific memory representations in the context of this task, this study may have actually underestimated the extent to which they would have otherwise produced evidence for audience design.

To examine these issues more directly, we carried out another study that investigated the influence of similar memory demands upon audience design (Horton and Gerrig 2005b). This study involved triads of participants as well, with one Director working in alternation with each of two Matchers on yet another version of the card-matching task. As in our earlier study, we varied Directors' initial experiences with each Matcher by giving the Matchers different subsets of cards, which in this study all contained categories of living things. For half of the triads, each Matcher worked initially with completely different card categories. For example, the Director might match sets of *fish*, *frogs*, *dogs*, and *lizards* with one Matcher, and *birds*, *cats*, *snakes*, and *flowers* with the other Matcher. We called this the *orthogonal* card condition. For the other triads, both Matchers dealt with items from all eight card categories, although the specific category exemplars were initially different across Matchers (e.g., two of the four cards in each category were given to one Matcher and the other two cards from the same categories given to the other Matcher). We called this the *overlapping* card condition. Because the orthogonal condition confounded individual Matchers with unique card categories, we expected it to enhance Directors' capacity to encode Matcher-specific information in memory. For the overlapping condition, however, both Matchers could be initially associated with the same card categories. Therefore, we expected Directors to have a harder time encoding (and subsequently using) Matcher-specific information in this condition. Any such encoding would have to be carried out on an item-by-item basis.

Subsequently, to test our claims we asked Directors to describe the full set of cards one final round with each Matcher. As in Horton and Gerrig (2002), half of the cards were unfamiliar to each Matcher, although all of the cards were highly familiar to the Director. The initial question then, was whether Directors' descriptions would be sensitive to whether a given card was familiar or unfamiliar to a particular Matcher. In general, we found evidence consistent with audience design. For cards unfamiliar to the current Matcher, Directors produced longer descriptions and were more likely to modify previous descriptions by providing additional information. However, in line with our predictions about the memory demands of each initial encoding condition, these effects were strongest for Directors who had initially experienced the orthogonal distribution of cards. This demonstrated the importance of considering the memory requirements of particular conversational situations when making predictions about the likelihood of observing audience design. Situations that foster the encoding of partner-specific information will enhance the capacity of speakers to subsequently use that information in ways consistent with audience design. Conversely, audience design may be less likely in situations that make partner-specific memory encoding more difficult.

6. Speaker-specific variation in memory-based process of audience design

Horton and Gerrig (2005) provided a demonstration of the manner in which aspects of discourse situations can influence the availability of memory representations relevant for audience design, and for message formation in particular. Memory availability, however, can vary across individuals as well. Simply put, some individuals may be better able to encode and retrieve relevant partner-specific memory representations in the time course necessary to have an influence upon language use. Recall that Horton and Keysar (1996) found that imposing time pressure on speakers reduced the extent to which their utterances showed sensitivity to the information available to their addressees (see also Roßnagel, 2000). Similarly, individuals that experience systemic difficulties in their ability to efficiently retrieve partner-relevant knowledge from memory may be less able to demonstrate evidence for audience design

One population that has been particularly well-studied with respect to differences in memory performance is elderly adults. Although a variety of changes in memory function can emerge as a consequence of normal aging, one particular aspect of memory that is potentially important for considerations of conversational common ground is *source memory*, which involves the recollection of the context or situation in which specific information was previously encountered (e.g., Johnson, Hashtroudi, and Lindsay 1993). A number of research findings has shown that older adults frequently have specific difficulties with memory for source information, even in situations when recollection of individual items is relatively good (e.g., Brown, Jones, and Davis, 1995; Chalfonte and Johnson 1996; Hashtroudi, Johnson, and Chrosniak 1989; Schacter et al. 1991). For example, Chalfonte and Johnson (1996) presented younger and older participants with different colored objects at different locations in an array, and found that older adults were impaired in recognizing particular combinations of objects and locations, although their recognition of contextual information in isolation was still relatively good. Chalfonte and Johnson proposed that older adults may have particular difficulty "binding" information about contextual details together with information about focal aspects of experiences.

These patterns of age-related changes in memory for associations and for source information are potentially informative about role of domain-general memory processes in conversational common ground. Given the memory-based view outlined above, which assumes that effects attributable to common ground emerge on the basis of the associations that people have with respect to other individuals, any impairment in the ability to encode or retrieve such associations should also impair audience design. If assumptions about common ground are built in part upon on the contextual associations that people have with respect to other individuals, then older adults' underlying difficulties encoding or retrieving such associations may also impair their ability to show evidence for audience design in message formation.

Prior research paints a mixed picture of the ability of older adults to adjust utterances for particular addressees. Some findings suggest that older speakers can tailor aspects of their speech based on general partner characteristics. For example, Gould and Shaleen (1999) found that elderly women could modify particular high-level aspects of their interactions, such as turn-taking and question-asking, when talking with a college student versus an individual with mild mental retardation. Similarly, Adams, Smith, Pasupathi, and Vitolo (2002) found that older women were more likely than young adults to simplify narratives when speaking to children. These results suggest that older adults can make relatively global adjustments to their speech when necessary.

Other evidence suggests that the speech of older adults may not always show clear evidence for partner-related adjustments. Kogan and Jordan (1989) found that elderly adults did little to vary the amount of their speech when talking to another elderly partner versus a middle-aged partner, although they did show more idiomatic or "personalized" features when talking to peers. An examination of task-oriented dialogue in pairs of younger adults, pairs of older adults, and mixed-age pairs found that older adults were less likely to vary the fluency, complexity, and content of their speech across partners compared to younger speakers (Kemper et al. 1995).

Additionally, findings from several referential communication studies have shown that pairs of older adults consistently require more time to find mutually acceptable referring expressions compared to pairs of younger adults (Bortfeld et al. 2001; Horton and Spieler 2007; Hupet, Chantraine, and Nef 1993). Hupet et al. (1993) also reported that older adults in these tasks were more likely to produce idiosyncratic descriptions that failed to incorporate previously established referring expressions, (although this may interact with the familiarity of the referential domain; cf. Horton and Spieler 2007). Taken together, these results suggest that older speakers may find it more difficult to consider specific knowledge about their partners as a routine aspect of message planning and production.

To explore this possibility more directly, Daniel Spieler and I carried out a study investigating the extent to which younger and older speakers would show evidence for audience design when placed in a communicative context that necessitated drawing upon partner-specific information from memory (Horton and Spieler 2007). Our study involved two task phases: an initial "familiarization" phase in which pairs of younger and pairs of older adults took part in a standard card-matching task, and a subsequent "picture description" phase in which the same participants, working as individuals, described pictures for each of two addressees. In typical referential communication fashion, the familiarization phase gave pairs of participants the opportunity to encode into common ground mutually-agreed upon ways of referring to the items in the card sets, which depicted different categories of living things. After this card-matching task, we separated the members of each pair and asked each person to describe the same items again in the context of a computer-mediated communication task. On each trial of this task, we presented via computer four items from one of the eight card categories (e.g., four pictures of cats), and the task was to describe a specific "target" in such a way that an addressee would be able to select the same picture on another computer. Although we simply recorded these descriptions via microphone, our cover story explained that the participants' utterances were being transmitted in real time to each of two possible addressees: the *familiar* partner with whom they had just completed the card-matching task or a completely naïve *unfamiliar* partner. The particular addressee supposedly responsible for selecting the target picture varied from trial to trial, and we conveyed this by presenting a digital image of the relevant addressee prior to each trial.

Because the picture descriptions involved the same items that had been present for the initial card-matching task, each participant shared relevant common ground for these items with the familiar addressee but not with the unfamiliar addressee. This allowed us to compare speakers' utterances across addressees for evidence of audience design, which we assessed through converging evidence from several measures: the number of words in speakers' descriptions, the speed with which they initiated these descriptions, and the extent to which their descriptions repeated information that had been established for the same items earlier, during the card-matching task. If participants were sensitive to the common ground shared with each addressee – and commonality assessment should not have been an issue given that we made it very clear which partner was the addressee on each trial – we expected that their descriptions for the familiar addressee would be shorter, be initiated more quickly, and be more similar to previous descriptions than their descriptions for the unfamiliar addressee.

Given our interest in the influence of memory-based processes on audience design, we placed two additional constraints on how speakers carried out the picture description task. First, communication was essentially one-way. That is, participants received no feedback from their simulated partners other than a random signal that supposedly indicated when the addressee had made a response. In typical conversational settings, addressees often provide immediate feedback when speakers produce confusing or incomplete utterances (Kraut, Lewis, and Swezey 1982). The expectation of that such feedback will be available may cause speakers to be relatively less careful with respect to audience design compared to situations when interaction is allowed. Because we designed our experimental context to preclude the possibility of feedback, we anticipated that speakers would be forced to rely more directly on their own best evaluations of how to describe items for each partner. Second, we limited the amount of time speakers were given to produce their descriptions. Based on pilot testing, younger adults were given 10 seconds and older adults were given 12.5 seconds to describe the target picture on each trial. When feedback is not available, speakers often fail to shorten their referring expressions over time as much as they would otherwise in more interactive situations (e.g., Hupet and Chantraine 1992; Krauss and Weinheimer 1966; Murfitt and McAllister 2001). By placing a time limit on participants in our task, we hoped to encourage them to rely on assessments of shared knowledge whenever possible. Because, however, we did not want to pressure the speakers unduly, the time limit was still intended to be relatively generous for both younger and older adults.

The results were quite consistent. Young adult speakers showed clear evidence for audience design; descriptions for familiar addresses were shorter, initiated more quickly, and more similar to previously established descriptions compared to descriptions for unfamiliar addressees.   The descriptions by older speakers, however, showed no sensitivity to the status of the addressees in any of these measures.   In general, the older adults appeared to treat the description task as if everything were "new" for both partners, producing relatively long, idiosyncratic descriptions in both contexts.   Notably, this flies in the face of their own self-reports; during post-experimental debriefing, the majority of the older adults reported being aware of the difference across partners and of the need to provide additional help for the unfamiliar addressee.   Even so, there was no evidence of any partner-specific adjustments in their actual performance on the task.

To what extent did these differences in task performance emerge from age-related differences in partner-specific memory encoding or retrieval? To examine this question, we examined a subset of older speakers who showed the sharpest gains in "efficient" performance on the initial card-matching task.   In general, individuals perform more efficiently over time on repeated referential communication tasks to the extent that they are able to encode into memory information about their partners' perspectives (Nohara-LeClair 2001).   If the generally poor performance on our picture description task were an encoding problem, then we reasoned that the older adults who showed the strongest evidence of successful initial encoding would be most likely to show evidence for audience design subsequently. This was not the case – picture descriptions produced by this subset of older adults showed no more sensitivity to the communicative needs of each addressee than those produced by the group as a whole – which suggested to us that the observed age-related differences in audience design were more likely due to difficulties with successfully retrieving and incorporating partner-related information at the time of message formation.

Although we had every reason to believe, based on initial screening, that the older adults in this study were healthy and cognitively intact for their age, an important caveat is that we were unable to systematically collect independent measures of the cognitive capabilities on our participant populations.   In order to tease apart the specific memory components relevant for different aspects of conversational phenomena like audience design, it will be necessary to obtain more specific information about possible variation in memory abilities.   Some intriguing evidence in this direction was recently reported by Duff, Hengst, Tranel, and Cohen (2006), who examined conversational collaborations in individuals suffering from an especially extreme memory impairment: hippocampal amnesia. In this study, a group of amnesic patients worked with partners on a standard referential communication task. The patients all exhibited severe impairments as measured by standardized memory tests such as the Wechsler Memory Scale-III. These memory deficits were due to focal lesions on the hippocampus, a subcortical brain structure heavily involved in encoding and retrieval of explicit, declarative memory (Squire 1992). To examine whether these impairments in explicit memory function would affect how individuals established and used common ground in conversation, Duff et al. asked four patients and four normal controls to serve as Director for multiple trials of a collaborative tangram-matching task together carried out in conjunction with a "familiar" partner (e.g., a spouse). Importantly, because the repeated trials were spread out over the course of two days, the explicit memory requirements of having to recollect what particular cards had been called in previous sessions in theory were beyond the capacities of the amnesic patients.

However, even though the patients were generally less efficient at working with their partners to match the cards compared to the normal controls–they required more time and more words to complete the task–they showed gains in efficiency over the course of the repeated trials that were remarkably similar to the patterns of performance improvements demonstrated by the controls. Moreover, when asked six months later to recall the labels for the Tangram figures that they had grounded with their partners, the amnesic patients were highly accurate, producing the correct labels over 80% of the time, the same rate as the normal controls. This was in marked contrast to the general inability of the amnesic patients to successfully learn *arbitrary* (i.e., non-negotiated) figure-label pairings for the same figures, consistent with other evidence suggesting that amnesia prevents certain forms of associative learning (Cohen and Eichenbaum 1993).

How can one explain the amnesic patients' impressive performance on this collaborative card-matching task? At the very least, these results argue against the view that successful management of conversational common ground necessarily requires consultation of some sort of explicit record of one's interactions with others. Access to any sort of continuously updated partner-model should have been beyond the memory capacities of these individuals. Duff et al. (2006) suggested that the collaborative matching task may have fostered a process of perceptual and conceptual convergence in the participant pairs such that particular Tangrams simply came to "look like" a man taking a siesta, for example. Neuropsychological research has shown that that this kind of perceptual learning is relatively spared in cases of hippocampal damage (Manns and Squire 2001). On this explanation, rather than explicitly recalling which names had been negotiated with their partners previously, the amnesic patients' reliance on implicit learning mechanisms enabled their use of increasingly consistent labels over the course of the interactions, leading to rates of collaborative improvement similar to those shown by the normal controls.

It would have been useful, of course, to know whether the conversational partners working with these amnesic patients provided cues, either explicitly or implicitly, that allowed the patients to more efficiently carry out the matching task. It is possible that the partners, who were all familiar with these patients, may have known how to provide feedback to the amnesic patients in ways that could partially compensate for difficulties with memory in conversation (Hengst 2003). Alternatively, however, the mere presence of the partners, together with the perceptual availability of the task stimuli, could have served as an implicit compound cue (McKoon and Ratcliff 1992) for the retrieval of associated information from memory. This would be consistent with the results of Horton (2007) and the present description of automatic commonality assessment as being rooted in implicit memory. Even so, the fact that the patients generally were able to identify the figures with the same labels up to 6 months later suggests that the figures by themselves were sufficiently good cues for the implicit retrieval of previously-established perspectives. The critical detail, of course, is that this retrieval must have taken place in the absence of any explicit memory representation of the task interactions.

7. Conclusions

In sum, then, this memory-based account of conversational common ground suggests that an important reason why language users are able to show evidence for audience design under many circumstances is because they rely upon ordinary mechanisms of memory encoding and retrieval for the purposes of commonality assessment and message formation (Horton and Gerrig 2005a). As domain-general memory processes, these mechanisms can support audience design to the extent that conversational situations foster the effective encoding and retrieval of partner-specific memory cues, and also to the extent that individual speakers have the capacity to encode and retrieve available partner-specific cues within a time course to have an impact upon utterance planning. Importantly, automatic processes of implicit memory – specifically, low-level resonance mechanisms that increase the accessibility of relevant memory traces – mediate the ability of language users to take partner-relevant information into account, without the need for an explicit model of the other's knowledge. When automatic processes of memory retrieval fail to return appropriate information, errors like "X-500 directories" are likely to occur.

As I have outlined in this chapter, this memory-based account incorporates particular features of existing models of common ground inference and of conversational coordination more generally. Like the interactive alignment account proposed by Pickering and Garrod (2004), this view sees important aspects of interpersonal interaction as being mediated through relatively automatic processes of priming and memory retrieval. Given the way these processes function much of the time, language use may happen opportunistically, based on the cognitive representations become most immediately accessible through these low-level mechanisms. When the representations that become most immediately available are primarily one's own, this opportunism can appear egocentric, as described by the perspective adjustment view (Barr and Keysar, 2006). However, when the cues in particular situations are such that partner-specific representations become strongly accessible in time to influence message planning and production, it should be possible to observe effects due to assessments of shared knowledge from the earliest moments of processing, consistent with constraint-based accounts of common ground inference (Hanna et al. 2003; Nadig and Sedivy 2002).

In general, I have tried to make it clear how the memory-based account defines the circumstances under which language use may or may not be expected to involve specific representations or processes dedicated to the problem of deriving beliefs about common ground. This is not to say that explicit considerations of shared knowledge never happen during conversation. Rather, the claim is simply that conversational phenomena like audience design can, in many circumstances, be mediated through domain-general memory processes. Indeed, there are many situations in which relatively strategic considerations of commonality would be expected to occur, either because of the need to keep track of what information is shared or not, or because feedback from the partner triggers the need for possible monitoring and error correction. Those moments in which language users devote effort toward partner-specific memory retrieval are undoubtedly similar to the types of accommodations for partners' perspectives described by the perspective adjustment view (Barr and Keysar 2006; Keysar et al. 2000).

Too often, past considerations of conversational pragmatics have neglected the fact that even complex aspects of language processing and use rest upon a foundation of more basic cognitive psychological mechanisms. The research presented here has demonstrated how psychologically plausible models of common ground will benefit from a better understanding of how conversational coordination between individuals can be affected by the cognitive processes at work within each individual separately. While it is highly unlikely that relatively simple processes like resonance and context-dependent memory will ever completely account on their own for the intricacies of reference generation and audience design, these processes are one important means through which speakers accommodate to others in conversation.

References

Albrecht, Jason E., and Jerome L. Myers
    1998.   Accessing distant text information during reading: Effects of
            contextual cues. *Discourse Processes* 26: 67–86.
Adams, Cynthia, Malcolm C. Smith, Monisha Pasupathi, and Loretta Vitolo.
    2002   Social context effects on story recall in older and younger women:
            Does the listener make a difference? *Journal of Gerontology:*
            Psychological Sciences 57: 28–40.
Barr, Dale J., and Boaz Keysar
    2002   Anchoring comprehension in linguistic precedents. *Journal of*
            Memory and Language 46: 391–418.
    2006   Perspective taking and the coordination of meaning in language use.
            In *Handbook of psycholinguistics*. 2d ed., Matthew J. Traxler and
            Morton A. Gernsbacher (eds.), 901–938. Amsterdam: Academic
            Press.
Bock, J. Kathryn
    1986   Syntactic persistence in language production. *Cognitive Psychology*
            18: 355–387.

Bortfeld, Heather, Silvia D. Leon, Jonathan E. Bloom, Michael F. Schober, and
      Susan E. Brennan
  2001   Disfluency rates in conversation: Effects of age, relationship, topic,
           role, and gender. *Language & Speech* 44: 123–147.
Branigan, Holly P., Martin J. Pickering, and Alexandra A. Cleland
  2000   Syntactic coordination in dialogue. *Cognition* 75: B13–B25.
Brennan, Susan E., and Herbert H. Clark
  1996   Conceptual pacts and lexical choice in conversation. *Journal of
           Experimental Psychology: Learning, Memory, and Cognition* 22:
           1482–1493.
Brown, Alan S., Elizabeth M. Jones, and Trina L. Davis
  1995   Age differences in conversational source monitoring. *Psychology
           and Aging* 10: 111–122.
Chalfonte, Barbara L., and Marcia K. Johnson
  1996   Feature memory and binding in young and older adults. *Memory &
           Cognition* 24: 403–416.
Chun, Marvin M., and Yuhong Jiang
  2003   Implicit, long-term spatial contextual memory. *Journal of
           Experimental Psychology: Learning, Memory, and Cognition* 29:
           224–234.
Clark, Herbert H.
  1992   *Arenas of language use.* Chicago: University of Chicago Press.
  1996   *Using language.* Cambridge: Cambridge University Press.
Clark, Herbert H., and Catherine R. Marshall
  1978   Reference diaries. In *TINLAP-2: Theoretical issues in natural
           language processing*-2, David L. Waltz (ed.), 57–63. New York:
           Association for Computing Machinery.
  1981   Definite reference and mutual knowledge. In *Elements of discourse
           understanding*, Aravind K. Joshi, Bonnie L. Webber, and Ivan A.
           Sag (eds.), 10–63. Cambridge: Cambridge University Press.
Clark, Herbert H., and G.L. Murphy
  1982   Audience design in meaning and reference. In *Language and
           comprehension,* J.F. Le Ny and Walter Kintsch (eds.), 287–299.
           Amsterdam: North-Holland.
Clark, Herbert H., Robert Schreuder, and Samuel Buttrick
  1983   Common ground and the understanding of demonstrative reference.
           Journal of Verbal Learning and Verbal Behavior 22: 245–258.
Clark, Herbert H., and Deanna Wilkes-Gibbs
  1986   Referring as a collaborative process. *Cognition* 22: 1–39.
Cohen, Neal J., and Howard Eichenbaum
  1993   *Memory, amnesia and the hippocampal system.* Cambridge, MA:
           MIT Press.

Dell, Gary S., and Paula M. Brown
   1991   Mechanisms for listener-adaptation in language production: Limiting
          the role of the "model of the listener." In *Bridges between
          psychology and linguistics: A Swarthmore festschrift for Lila
          Gleitman*, Donna J. Napoli and Judy A. Kegl (eds.), 105–129.
          Hillsdale, NJ: Erlbaum.
Duff, Melissa C., Julie Hengst, Daniel Tranel, and Neal J. Cohen
   2006   Development of shared information in communication despite
          hippocampal amnesia. *Nature Neuroscience* 9: 140–146.
Gerrig, Richard J., and Gail McKoon
   1998   The readiness is all: The functionality of memory-based text
          processing. *Discourse Processes* 26: 67–86.
Gerrig, Richard J., and Edward J. O'Brien
   2005   The scope of memory-based processing. *Discourse Processes*
          39:225–242.
Gould, Odette N., and Lori Shaleen
   1999   Collaboration with diverse partners: How older women adapt their
          speech. *Journal of Language and Social Psychology* 18: 395–418.
Grice, H. Paul
   1975   Logic and conversation. In *Syntax and semantics: Speech acts, Vol.
          3*, Peter Cole and Jerry L. Morgan (eds.), 41–58. New York:
          Academic Press.
Gillund, Gary, and Richard M. Schiffrin
   1984   A retrieval model for both recognition and recall. *Psychological
          Review* 91: 1–67.
Hanna, Joy E., Michael K. Tanenhaus, and John Trueswell
   2003   The effects of common ground and perspective on domains of
          referential interpretation. *Journal of Memory and Language* 49: 43–
          61.
Hengst, Julie
   2003   Collaborative referencing between individuals with aphasia and
          routine communication partners. *Journal of Speech, Language, and
          Hearing Research* 46: 831–848.
Hintzman, Douglas L.
   1986   "Schema abstraction" in a multiple-trace memory model.
          Psychological Review 93: 411–428.
Horton, William S.
   2007   The influence of partner-specific memory associations on language
          production: Evidence from picture naming. *Language and Cognitive
          Processes* 22: 1114–1139.

Horton, William S., and Richard J. Gerrig
   2002   Speakers' experiences and audience design: Knowing *when* and
            knowing *how* to adjust utterances to addressees. *Journal of Memory
            and Language* 47: 589–606.
   2005a   Conversational common ground and memory processes in language
            production. *Discourse Processes* 40: 1–35.
   2005b   The impact of memory demands on audience design during language
            production. *Cognition* 96: 127–142.
Horton, William S., and Boaz Keysar
   1996   When do speakers take into account common ground? *Cognition*
            59: 91–117.
Horton, William S., and Daniel H. Spieler
   2007   Age-related effects in communication and audience design.
            Psychology and Aging 22: 281–290.
Hupet, Michel, and Yves Chantraine
   1992   Changes in repeated references: Collaboration or repetition effects?
            Journal of Psycholinguistic Research 21: 485–496.
Hupet, Michel, Yves Chantraine and F. Nef
   1993   References in conversation between young and old normal adults.
            Psychology and Aging 8: 339–346.
Johnson, Marcia K., Shahin Hashtroudi, and D. Stephen Lindsay
   1993   Source monitoring. *Psychological Bulletin* 114: 3–28.
Kemper, Susan
   1994   Elderspeak: Speech accommodations to older adults. *Aging and
            Cognition* 1: 17–28.
Kemper, Susan, Dixie Vandeputte,Karla Rice, Him Cheung, and Julia Gubarchek
   1995   Speech adjustments to aging during a referential communication
            task. *Journal of Language and Social Psychology* 14: 40–59.
Keysar, Boaz, Dale J. Barr, Jennifer A. Balin, and Jason S. Brauner
   2000   Taking perspective in conversation: The role of mutual knowledge in
            comprehension. *Psychological Science* 11: 32–38.
Keysar, Boaz, Dale J. Barr, Jennifer A. Balin, and Timothy S. Paek
   1998   Definite reference and mutual knowledge: Process models of
            common ground in comprehension. *Journal of Memory and
            Language* 39: 1–20.
Keysar, Boaz, Dale J. Barr, and William S. Horton
   1998   The egocentric bias of language use: Insights from a processing
            approach. *Current Directions in Psychological Science* 7 : 46–50.
Kingsbury, Paul, Stephanie Strassel, Cynthia McLemore, and Robert McIntyre
   1997   CALLHOME American English transcripts, LDC97T14.
            Philadelphia: Linguistic Data Consortium.

Kogan, Nathan, and T. Jordan
   1989   Social communication within and between elderly and middle-aged
           cohorts. In *Psychological development: Perspectives across the life-
           span*, Mary A. Luszcz and Ted Nettelbeck (eds.), 409–415.
           Amsterdam: North-Holland.
Kraut, Robert E., Steven H. Lewis,and Lawrence W. Swezey
   1982   Listener responsiveness and the coordination of conversation.
           Journal of Personality and Social Psychology 43: 718–731.
Krauss, Robert M., and Sidney Weinheimer
   1964   Changes in reference phrases as a function of frequency of usage in
           social interaction: A preliminary study. *Psychonomic Science* 1:
           113–114.
   1966   Concurrent feedback, confirmation, and the encoding of referents in
           verbal communication. *Journal of Personality and Social
           Psychology* 4: 343–346.
Lea, R. Brooke, Robert A.Mason, Jason E. Albrecht, Stacy L. Birch, and Jerome L.
           Myers
   1998   Who knows what about whom: What role does common ground play
           in accessing distant information? *Journal of Memory and Language*
           39: 70–84.
Lee, Benny P. H.
   1999   Mutual knowledge, background knowledge and shared beliefs: Their
           roles in establishing common ground. *Journal of Pragmatics* 33: 21–
           44.
Lewis, David K.
   1969   *Convention: A philosophical study*. Cambridge, MA: Harvard
           University Press.
MacDonald, Maryellen C., Neal J. Pearlmutter, and Mark S. Seidenberg
   1994   Lexical nature of syntactic ambiguity resolution. *Psychological
           Review* 101: 676–703.
Manns, Joseph R., and Larry R. Squire
   2001   Perceptual learning, awareness, and the hippocampus. *Hippocampus*
           11: 776–782.
McKoon, Gale, and Roger Ratcliff
   1979   Priming in episodic and semantic memory. *Journal of Verbal
           Learning and Verbal Behavior* 18: 463–480.
   1992   Spreading activation versus compound cue accounts of priming:
           Mediated priming revisited. *Journal of Experimental Psychology:
           Learning, Memory, and Cognition* 18: 1155–1172.
McRae, Ken, Michael J. Spivey-Knowlton and K. Michael Tanenhaus
   1998   Modeling the influence of thematic fit (and other constraints) in on-
           line sentence comprehension. *Journal of Memory and Language* 38:
           283–312.

Murfitt, Tara, and Jan McAllister
2001   The effect of production variables in monolog and dialog on comprehension by novel listeners. *Language and Speech* 44: 325–350.
Nadig, Aparna S., and Julie C. Sedivy.
2002   Evidence of perspective-taking constraints in children's on-line reference resolution. *Psychological Science* 13 : 329–336.
Nohara-LeClair, Michiko
2001   A direct assessment of the relation between shared knowledge and communication in a referential communication task. *Language and Speech* 44: 217–236.
Olson, David R.
1970   Language and thought: Aspects of a cognitive theory of semantics. *Psychological Review* 77: 257–273.
Parker, Amanda, Angus Gellatly, and Mitchell Waterman
1999   The effect of environmental context manipulation on memory: Dissociation between perceptual and conceptual implicit tests. *European Journal of Cognitive Psychology* 11: 555–570.
Pecher, Diane, and Jeroen G.W. Raaijmakers
2004   Priming for new associations in animacy decision: Evidence for context dependency. *The Quarterly Journal of Experimental Psychology* 57A: 1211–1231.
Pickering, Martin J., and Simon Garrod
2004   Toward a mechanistic psychology of dialogue. *Behavioral and Brain Sciences* 27: 169–226.
2006   Alignment as the basis for successful communication. *Research on Language and Computation* 4: 203–228.
Ratcliff, Roger
1978   A theory of memory retrieval. *Psychological Review* 85: 59–108.
Roßnagel, Christian
2000   Cognitive load and perspective taking: Applying the automatic-controlled distinction to verbal communication. *European Journal of Social Psychology* 30: 429–445.
Schacter, Daniel L., Alfred W. Kaszniak, John F. Kihlstrom, and Michael Valdiserri
1991   The relation between source memory and aging. *Psychology and Aging* 6: 559–568.
Schiffer, Stephen R.
1972   *Meaning.* Oxford: Clarendon Press.
Schiffrin, Richard M., and Mark Steyvers
1997   A model for recognition memory: REM: Retrieving effectively from memory. *Psychonomic Bulletin and Review* 4: 145–166.

Schober, Michael F., and Susan E. Brennan
   2003   Processes of interactive spoken discourse: The role of the partner. In
             Handbook of discourse processes, Arthur C. Graesser, Morton A.
             Gernsbacher, and Susan R. Goldman (eds.), 123–164. Mahwah, NJ:
             Lawrence Erlbaum.
Smith, Neilson V. (ed.)
   1982   *Mutual knowledge*. London: Academic Press.
Smith, Steven M.
   1994   Theoretical principles of context-dependent memory. In *Theoretical
             aspects of memory*, 2d ed. Peter E. Morris and Michael Gruneberg
             (eds.), 168–195. London: Routledge.
Snow, Catherine E., and Charles A. Ferguson
   1977   *Talking to children: Language input and acquisition*. Cambridge:
             Cambridge University Press.

Spieler, Daniel H., and David A. Balota
   1996   Characteristics of associative learning in younger and older adults:
             Evidence from an episodic priming paradigm. *Psychology and
             Aging* 11: 607–620.
Squire, Lawrence R.
   1992   Memory and the hippocampus: A synthesis from findings with rats,
             monkeys, and humans. *Psychological Review* 99: 195–231.
Tulving, Endel
   1983   *Elements of episodic memory*. Oxford: Oxford University Press.
   2002   Episodic memory: From mind to brain. *Annual Review of
             Psychology* 53: 1–25.

Common ground as a resource for social affiliation

Nicholas J. Enfield

1. Introduction

The pursuit and exploitation of mutual knowledge, shared expectations, and other types of common ground (Clark 1996; Lewis 1969; Smith 1980)[1] not only serves the mutual management of referential information, but has important consequences in the realm of social, interpersonal affiliation. The informational and social-affiliational functions of common ground are closely interlinked. I shall argue in this chapter that the management of information in communication is never without social consequence, and that many of the details of communicative practice are therefore dedicated to the management of social affiliation in human relationships.

Common ground constitutes the open stockpile of shared presumption that fuels amplicative inference in communication (Grice 1989), driven by intention attribution and other defining components of the interaction engine (Levinson 1995, 2000, 2006). Any occasion of "grounding" (i.e., any increment of common ground) has consequences for future interaction of the individuals involved, thanks to two perpetually active imperatives for individuals in social interaction. An *informational imperative* compels individuals to cooperate with their interactional partners in maintaining a common referential understanding, mutually calibrated at each step of an interaction's progression. Here, common ground affords economy of expression. The greater our common ground, the less effort we have to expend to satisfy an informational imperative. Second (but not secondary), an *affiliational imperative* compels interlocutors to maintain a common degree of interpersonal affiliation (trust, commitment, intimacy), proper to the status of the relationship, and again mutually calibrated at each step of an interaction's progression. In this second dimension, the economy of expression enabled by common ground affords a public display of intimacy, a reliable indicator of how much is personally shared by a given pair (trio, n-tuple) of interactants. In these two ways, serving the ends of informational economy and affiliational intimacy, to increment common ground is to invest in a resource that will be drawn on later, with interest.

2. Sources of common ground

A canonical source of common ground is joint attention, a unique human practice that fuses perception and inferential cognition (Moore and Dunham 1995; Tomasello 1999, 2006). In joint attention, two or more people simultaneously attend to a single external stimulus, together, each conscious that the experience is shared. Fig. 1 illustrates a typical, everyday joint attentional scene.

Figure 1.

In this example, the fact that a washing machine is standing in front of these women is incontrovertibly in common ground thanks both to its physical position in the perceptual field of both interactants and to its operating panel being the target of joint attentional hand gestures (Kita 2003; Liszkowski 2006). But common ground is also there when it is not being signaled or otherwise manifest directly. At a personal level, the shared experiences of interactants are in common ground as long as the interactants know (and remember!) they were shared. At a cultural level, common ground may be indexed by signs of ethnic identity, and the common cultural background such signs may entail. One such marker is native dialect (as signaled, e.g., by accent), a readily detectable and reliable indicator of long years of common social and cultural experience (Nettle and Dunbar 1997; Nettle 1999). Suppose I begin a conversational exchange with a stranger of similar age to myself, who, like me, is a native speaker of Australian English. We will each immediately recognize this common native origin from each other's speech, and then I can be pretty sure that my new interlocutor and I will share vast cultural common ground from at least the core years of our linguistic and cultural socialization (i.e., our childhoods, when our dialects were acquired). We will mutually assume, for instance, recognition of expressions like *fair dinkum*, names like Barry Crocker, and possibly even sporting institutions like the Dapto Dogs.

3. Common ground as fuel for Gricean ampliative inference

Common ground is a resource that speakers exploit in inviting and deriving pragmatic inference, as a way to cut costs of speech production by leaving much to be inferred by the listener. As Levinson (2000) points out, the rate of transfer of coded information in speech is slow, thanks to our articulatory apparatus. Psychological processes run much faster. This bottleneck problem is solved by the ampliative properties of pragmatic inference (Levinson 2000; cf. Grice 1975). Interpretative amplification of coded messages feeds directly on the stock of common ground, in which we may include a language's semantically coded linguistic categories (lexicon, morphosyntax), a community's set of cultural practices and norms (Levinson 1995: 240; Enfield 2002: 234–236), and shared personal experience. (This implies different categories of social relationship, defined in part by amount and type of common ground: e.g., speakers of our language, people of our culture, and personal associates of various types; see below.) This logic of communicative economy—intention attribution via inference fed by common ground—is complemented by the use of convention to simplify problems of social coordination (Clark 1996; Lewis 1969; Schelling 1960). Although we have access at all times to the powerful higher-order reasoning that makes common ground and intention attribution possible, we keep cognition frugal by assuming defaults where possible (Gigerenzer et al. 1999; Sperber and Wilson 1995; cf. Barr and Keysar 2004). So, if tomorrow is our weekly appointment (midday, Joe's) we do not have to discuss where and when to meet. The hypothesis that we will meet at Joe's at midday has been tested before, [2] and confirmed. And we further entrench the convention by behaving in accordance with it (i.e., by turning up at Joe's at midday and finding each other there).

Consider a simple example from everyday interaction in rural Laos, which illustrates common ground from both natural and cultural sources playing a role in inference making. Fig. 2 is from a video recording of conversation among speakers of Lao in a lowland village near Vientiane, Laos. (The corners of the image are obscured by a lens hood.) The image shows a woman (foreground, right; hereafter, Foreground Woman [FW]) who has just finished a complex series of preparations to chew betel nut, involving various ingredients and tools kept in the basket visible in the lower foreground. In this frame, FW is shifting back, mouth full with a betel nut package, having finished with the basket and placed it aside, to her left:

Figure 2.

Immediately after this, the woman in the background, at far right (Background Woman [BW]), moves forward, to reach in the direction of the basket, as shown in Fig. 3a, b.

Figure 3a.   *Figure 3b.*

BW's forward-reaching action gives rise to an inference by FW that BW wants the basket.[3] We can tell FW has made this inference from the fact that she grasps the basket and passes it to BW in Fig. 4. And we can tell, in addition, from what she says next, in line 1 of (1), that she infers BW wants to chew betel nut (the numbers at the end of each Lao word mark lexical tone distinctions):

Figure 4a.   *Figure 4b.*

<pre>
(1)   1 FW   caw4   khiaw4 vaa3
              2SG.P   chew   QPLR.INFER
              You chew?
      2 BW   mm5
              INTJ
              Mm. (i.e. <i>Yep.</i>)
</pre>

FW infers more than one thing from the forward-reaching action of BW shown in Fig. 3. It would seem hardly culture specific that BW is taken to want the basket. (But an inference or projection is nevertheless being made; after all, she may have wanted to rub a spot of dirt off the floor where the basket was sitting.) More specific to the common ground that comes with this cultural setting, BW's reaching for the basket is basis for an inference that she wants to chew betel nut (and not, for instance, that she wants to reorganize the contents of the basket, or tip it out, or put it away, or spit into it). The inference that BW wants to chew betel nut is made explicit in the proposition in line 1 "you chew." The added sentence-final "evidential interrogative" particle *vaa3* (Enfield 2007b:45) makes explicit, in addition, *that* it is an inference. The particle *vaa3* encodes the notion that an inference has been made, and seeks confirmation that this inference is correct: that is, in a sequence *X vaa3*, the meaning of *vaa3* can be paraphrased along the lines "Something makes me think X is the case, you should say something now to confirm this." BW responds appropriately with a minimal spoken confirmation in line 2.

The two inferences made in this example—one, that BW's forward movement indicates she wants to take hold of something in front of her, and two, that she wants to have the basket to chew betel nut—are launched from different types of categorical knowledge (though they are both based on the attribution of intention through recognition of an agent's "attitude"; Mead 1934, see also Kockelman 2005). The first is a general stock of typifications determined naturally, essentially by biology: naive physics, parsing of motor abilities (Byrne 2006), frames of interpretation of experience arising through terrestrial fate (Levinson 1997: 28). A second basis for inference is the set of categories learned in culture—here, from the fate of being born in a Lao-speaking community, and acquiring the frames, scripts, and scenarios (Schank and Abelson 1977) of betel-nut chewing among older ladies in rural Laos (e.g., that betel paraphernalia is "free goods" that any middle-aged or older woman may reach for in such a setting—had a man or a child made the same reaching action here, they would not have been taken to be embarking on a betel-chewing session). Both these types of knowledge are in the common ground of these interlocutors, in the strict sense of being information openly shared.

4.   Grounding for inferring: The informationally strategic pursuit of common ground

Links between joint attention, common ground, and pragmatic inference suggest a process of grounding for inferring, by which the requirements of human sociality direct us to tend—while socializing—to dimensions of common ground that may be exploited in later socializing.[4] This formulation highlights the temporality of the connection between grounding (i.e., securing common ground) and inferring. Grounding is an online process (enabled by joint attention). Later inferring based on common ground presupposes or indexes the earlier establishment of that common ground (or indexes a presumption of that common ground, based on some cue, such as a person's individual identity, or some badge of cultural or subcultural identity).

Grounding for inferring takes place at different levels of temporal grain—that is, with different time lags between the point of grounding and the point of drawing some inference based on that grounding. At a very local level, it is observable in the structure of reference management through discourse (Fox 1987). Canonically, a referent's first mention is done with a full noun phrase (e.g., a name or a descriptive reference), with subsequent mentions using a radically reduced form (such as a pronoun; recorded example from Fox 1987:20, transcription simplified):

(2)　A: Did they get rid of <u>Kuhleznik</u> yet?
　　　B: No in fact I know somebody who has <u>her</u> now.

Forms like *her* do not identify or describe their referent. Their reference must be retrieved by inference or other indexical means. This is straightforward when a full form for the antecedent is immediately prior, as in (2). But if you miss the initial reference, lacking the common ground required for inferring what *her* must be referring to, you might be lost. Without the benefit of informative hand gestures or other contextual cues, you are likely to have to disrupt the flow of talk by asking for grounding, to be able to make the required referential inferences.

At a step up in temporal distance between grounding and payoff are forward-looking "setups" in conversational interaction (Jefferson 1978; Sacks 1974; cf. Goodwin's "prospective indexicals"; Goodwin 1996: 384), which, for instance, alert listeners to the direction in which a speaker's narrative is heading. When I say "Her brother is so strange, let me tell you what he did last week," you as listener will then need to monitor my narrative for something that is sufficiently strange to count as the promised key illustration of her brother's strangeness, and thus the punch line. What constitutes "her brother's strangeness" is "not yet available to recipients but is instead something that has to be discovered subsequently as the interaction proceeds" (Goodwin 1996:384). When you hear what you think is this punch line, you will likely surmise that the story is at completion. Your response will be shaped by a second function of the prospective expression, namely, as a forewarning of the appropriate type of appraisal that the story seeks as a response or receipt. So, "He's so strange, let me tell you…" will rightly later elicit an appreciation that is fitted to the projected assessment; for example, *Wow, how weird.* Setup expressions of this kind are one type of grounding for inferring, with both structural-informational functions (putting in the open the fact that the speaker is engaged in a sustained and directed activity of telling—e.g., "how strange her brother is"), and social-affiliational functions (putting in the open the speaker's stance toward the narrated situation, which facilitates the production of affiliative, or at least fitted response). Both these functions help constrain a listener's subsequent interpretation as appropriate to the interaction, at a discourse level.

All the way at the other end of the scale in temporal distance between grounding and its payoff are those acts of building common ground that look ahead into the interactional future of the people involved. At a personal level, our efforts to maintain and build common ground have significant consequences for the type of relationship we succeed in ongoingly maintaining, that is, whether we are socially close or distant (see below). At a cultural level, in children's socialization we spend a lot of time explaining and acting out for children "what people do," "what people say," and "how things are." This builds the cultural common ground that will soon streamline an individual's passage through the moment-by-moment course of their social life.

5.   Semiotics: Cognition and perception, structure, and emergence

A matter of some contention is the degree of involvement of higher-order cognition in these social interactional processes. Despite currency of the term "mind reading" and its variants in literature on social intelligence (Baron-Cohen 1995; Carruthers and Smith 1996; Astington 2006), we cannot read each other's minds. Miller wrote, "One of the psychologist's great methodological difficulties is how he can make the events he wishes to study publicly observable, countable, measurable" (1951: 3). This problem for the psychologist is a problem for the layperson too. In interaction, normal people need, at some level, to be able to model each other's (evolving and contingent) goals, based solely on perceptible information, by attending to one another's communicative actions and displays (Mead 1934). A no-telepathy assumption means that there is "no influencing other minds without mediating artifactual structure" (Hutchins and Hazlehurst 1995). As a result, semiosis—the interplay of perception and cognition, rooted in ethology and blossoming in the modern human mind—is a cornerstone of human sociality (Kockelman 2005; Peirce 1965). Humans augment the ethologically broad base of iconic and indexical meaning with symbolic structures and higher-order processes of intention attribution.

So if action and perception are the glue in human interaction, higher-order cognition is the catalyst. I see this stance as a complement, not an alternative, to radically interactionist views of cognition (cf. Molder and

Potter 2005). Authors like Norman (1991), Hutchins (1995), and Goodwin (1994, 1996) are right to insist that the natural exercising of cognition is in distributed interaction with external artifacts. And we must add to these artifacts our bodies (Enfield 2005; Goodwin 2000; Hutchins and Palen 1993) and our social associates (Goodwin 2006). Similarly, the temporal-logical structures of our social interactions are necessarily collaborative in their achievement (Clark 1996; Schegloff 1982), as may be our very thought processes (Goody 1995; Mead 1934, Rogoff 1994; Vygotsky 1962). But as individuals, we each physically embody and transport with us the wherewithal to move from scene to scene and still make the right contributions. We store cognitive representations (whether propositional or embodied) of the conventional signs and structures of language, of the cultural stock of conventional typifications that allow us to recognize what is happening in our social world (Schutz 1970), and of more specific knowledge associated with our personal contacts. And we have the cognitive capacity to model other participants' states of mind as given interactions unfold (Mead 1934).

Accordingly, here is my rephrasing of Miller (with a debt to Schutz 1970 and Sacks 1992): *One of the man in the street's great methodological difficulties is how he can understand (and make himself understood to) his social associates solely on the basis of what is publicly observable.* Any model of multiparty interaction will have to show how the combination of a physical environment and a set of mobile agents will result in emergence of the structures of interactional organization that we observe. It will also have to include descriptions of the individual agents, their internal structure and local goals. General capacities of social intelligence, and specific values of common ground will have to be represented somewhere in those individual minds. Then, in real contexts, what is emergent can emerge.

So, human social interaction not only involves cognition, it involves high-grade social intelligence (Goody 1995; allowing that it need not *always* involve it—Barr and Keysar 2004). And in line with those who resist the overuse or even abuse of mentalistic talk in the analysis of social interaction, it is clear that intention attribution is entirely dependent on perception in a shared environment (see esp. Byrne 2006; Danziger 2006; Goodwin 2000, 2006; Hutchins 1995:ch. 9, 2006; Schegloff 1982:73). Both components–individual cognition and emergent organization–are absolutely necessary (see Enfield and Levinson 2006). Human social interaction would not exist as we know it without the cocktail of individual, higher-order cognition and situated, emergent, distributed organization. A mentalist stance need therefore not be at the expense of the critically important emergence of organization from collaborative action in shared physical context, above and beyond any individual's internally coded goals. To be sure, there remain major questions as to the relative contribution of individual cognition and situated collaborative action in causing the observed organization of interaction. But however you look at it, we need both.

6. Audience design

Equipped with higher-order inferential cognition, an interlocutor (plus all the other aspects of one's interactional context), and a stock of common ground, a speaker should design his or her utterances for that interlocutor (Clark 1996; Sacks 1992; Sacks and Schegloff 1979; Schegloff 1997; Enfield and Stivers 2007; Kitzinger and Lerner 2007). If we are to optimize the possibility of having our communicative intentions correctly recognized, any attempt to make the right inferences obvious to a hearer will have to take into account the common ground defined by the current speaker-hearer combination. In ordinary conversation, there is no generic, addressee-general, mode of message formulation. To get our communicative intentions recognized, we ought to do what we can to make them the most salient solutions to the interpretive problems we foist on our hearers. The right ways to achieve this will be determined in large part by what is in the common ground, and this is by definition a function of who is being addressed given who it is they are being addressed by. Because Gricean implicature is fundamentally audience driven (whereby formulation of an utterance is tailored by how one expects an addressee will receive it), to do audience design is to operate at a yet higher level than mere intention attribution. It entails advance modeling of *another's* intention attribution.[5]

Consider an example that turns on highly local common ground. Fig. 5 shows two men sitting inside a Lao village house, waiting while lunch is prepared in an outside kitchen.

Figure 5.

At the moment shown in Figure 5, a woman's voice can be heard (coming from the outside kitchen verandah, behind the camera, left of screen) as follows:

(3) mòòt4   nam4   haj5   nèè1
     extinguish   water give   IMP.SOFT
     Please turn off the water for (me).

In making this request, the speaker does not explicitly select an addressee. Anyone in earshot is a potential addressee. Within a second or two, the man on the left of frame gets up and walks to an inside wall of the house, where he flicks an electric switch.

Consider the mechanism by which the utterance in (3) brought about this man's compliance. Although the woman's call in (3) was not explicitly addressed to a particular individual, it was at the very least for someone who was in hearing range *and* knew what compliance with the request in (3) entailed. Although relative social rank of hearers may work to narrow down who is to carry out the request, it remains that the utterance in (3) could not be intended for someone who *lacks* the common ground, that is, who does not know what "turning off the water" involves. The switch that controls an outside water pump is situated at the only power outlet in the house, inside, far from the kitchen verandah. To respond appropriately to the utterance in (3), an addressee would need this inside knowledge of what "turning off the water" entails. Without it, one might not even realize that the addressee of (3) is someone (anyone) *inside* the house. But it is in the common ground for the people involved in this exchange. They are neighbors of this household, daily visitors to the house. The woman outside on the verandah knows that the people inside the house know (and know that they are known to know!) the routine of flicking that inside switch to turn the outside water pump on and off. This enables the success of the very lean communicative exchange consisting of the spoken utterance in (3) and the response in Fig. 6.

Figure 6a.   *Figure 6b.*

Much is inferred by the actor in Fig. 6 beyond what is encoded in the spoken message in (3), in the amplicative sense outlined above. In addition, this example illustrates a defining feature of common ground information, namely that people cannot deny possessing it.[6] The man on our left in Fig. 5–who is situated nearest the switch–might not feel like getting up, but he could not use as an excuse for inaction a claim that he does not know what the speaker in (3) wants (despite the fact that nothing in her utterance makes this explicit).

The principle of audience design dovetails with common ground, because both are defined by a social relationship between interlocutors. As prefigured above, the general imperative of audience design is served by two, more specific imperatives of conversation. I described one of these— the informational imperative—as the cooperative struggle to maintain common referential understanding, mutually calibrated at each step of an interaction's trajectory (Clark 1996; Schegloff 1992). This will be satisfied by various means including choice of language spoken, choice of words, grammatical constructions, gestures, and the various devices for meeting "system requirements" for online alignment in interaction (mechanisms for turn organization, signals of ongoing recipiency, correction of errors and other problems, etc.; Goffman 1981:14; Schegloff 2006). Less well understood are the "ritual" requirements of remedial face work, and the need to deal with "implications regarding the character of the actor and his evaluation of his listeners, as well as reflecting on the relationships between him and them" (Goffman 1981:21; cf. Goffman 1967, 1971). We turn now to those.

7. The affiliational imperative in social interaction

Any time one is engaged in social interaction, one's actions are of real consequence to the social relationship currently being exercised. If you are acting too distant, or too intimate, you are most likely going to be held accountable for it. Heritage and Atkinson (1984:6) write that there is "no escape or time out" from the considerations of interaction's sequential, contextual nature. Similarly, there is no escape or time out from the social-relational consequences of interaction. Just as each little choice we make in communicative interaction can be assessed for its optimality for information exchange, it can equally be assessed for its optimality for maintaining (or forging) the current social relationship at an appropriate level of intensity or intimacy. The management of common ground is directly implicated in our perpetual attendance to managing personal relationships within our social networks. Next, I elaborate some mechanisms by which this is achieved, but first I want to flesh out what is meant by degrees of intimacy or intensity in social relationships.

One of the key tasks of navigating social life is maintaining positions in social networks, where relationships between individuals are carried through time, often for years. There are logical constraints on the nature of an individual's network of relationships thanks to an inverse relationship between time spent interacting with any individual, and number of individuals with whom one interacts. We have only so much time in the day, and sustained relationships cannot be multiplied beyond a certain threshold (cf. grooming among primates; Dunbar 1993, 1996). Spending more time interacting with certain individuals means more opportunities to increment common ground with those individuals, by virtue of the greater opportunity to engage in joint attentional activity such as conversation. This results in greater access to amplicative inference in communication. A corollary is having less time to interact with *others,* and thus less chance to increment common ground through personal contact with those others, and, in turn, less potential to exploit ampliative inference in communication with them.

Such considerations of the logical dynamics of time and social group size have been taken to suggest inherent biases in the organization of social network structure (Dunbar 1998; Dunbar and Spoors 1995; Hill and Dunbar 2003). Hill and Dunbar suggest that social networks are "hierarchically differentiated, with larger numbers of progressively less intense relationships maintained at higher levels" (2003: 67; cf. Dunbar 1998). They propose a model with inclusive levels (Hill and Dunbar 2003: 68; note that they also discuss groupings at higher levels than this)

(4)	Level of relationship intensity	Approximate size of group
	support clique	7
	sympathy group	21
	band	35
	social group	150

What defines membership in one or other of these levels? As with physical grooming among primates, those who I spend more time with in committed engagement will tend to be those who I can later rely on in times of trouble (and, similarly, to whom I will be obliged to offer help if needed). In some societies this will be somewhat preordained (e.g., by kin or equivalently fixed social relations), whereas in other types of societies people may be more freely selective (as in many modern urban settings). For humans, unlike in primitive physical grooming, such rounds of engagement are intertwined with the deployment of delicate and sophisticated symbolic structure (language), and so it is not (just?) a matter of *how long* we spend interacting with whom, but of *what kind of information is traded and thereby invested in common_ground.* This is why in one type of society I might have a more intensive, closer relationship with my best friend, even though I see very much less of him than my day-to-day professional colleagues.

Cultures will differ with respect to the determination of relationship intensity (quantitatively and qualitatively defined), and the practices by which such intensity is maintained and signaled. Hill and Dunbar suggest that a hierarchical structure of social relatedness like (4), above, will be maintained in more or less any cultural setting, but the qualitative basis for distinction between these levels in any given culture will be "wholly open to negotiation" (i.e., by the traditions of that culture; 2003:69). They cite various types of social practice that may locally define the relevant level of relationship: those from whom we get our hair care (among the !Kung San; Sugawara 1984), those "whose death would be personally devastating" (Buys and Larson 1979), those "from whom one would seek advice, support, or help in times of severe emotional or financial stress" (Dunbar and Spoors 1995), those to whom we would send Christmas cards (Hill and Dunbar 2003; the other citations in this sentence are also from Hill and Dunbar 2003: 67). An important empirical project is the investigation of commonality and difference in how people of different cultures mark these social distinctions through interactional practice (regardless of whether membership in different levels of relationship intensity in a given setting is socioculturally predetermined or selected by individuals' preference).

Practices concerned with the management of common ground for strategic interactional purposes provide, I suggest, an important kind of data for assessing Hill and Dunbar's proposal. Given the "no time out" nature of everyday interaction, we may better look to practices that are very much more mundane and constant in the lives of regularly interacting individuals than, say, annual gestures like the Anglo Christmas Card. To this end, I want to draw a key link, so far entirely unseen in the literature, it seems, between the line of thinking exemplified by Hill and Dunbar (2003), and a strand of work arising from research within corners of sociology on conversation and other types of interaction, rooted in the work of Sacks and associates on "social membership categorization" (cf. Sacks 1992; see also Garfinkel and Sacks 1970; Schegloff 2002). In a review of this work, Pomerantz and Mandelbaum (2005) outline four types of practice in U.S. English conversation by which people "maintain incumbency in complementary relationship categories, such as friend–friend, intimate–intimate, father–son, by engaging in conduct regarded as appropriate for incumbents of the relationship category and by ratifying appropriate conduct when performed by the cointeractant" (Pomerantz and Mandelbaum 2005:160):

(5) Four sets of practices for maintaining incumbency in more intensive/intimate types of social relationships (derived from Pomerantz and Mandelbaum 2005):

— "Inquiring about tracked events and providing more details on one's own activities": reporting and updating on events and activities mentioned in previous conversations; eliciting detailed accounts, demonstrating special interest in the details; attending to each other's schedules and plans; and so forth (Drew and Chilton 2000; Morrison 1997).

— "Discussing one's own problems and displaying interest in the other's problems": claiming the right to (and being obliged to) ask and display interest in each other's personal problems; showing receptivity to such discussion; and so forth (Cohen 1999; Jefferson and Lee 1980).

— "Making oblique references to shared experiences and forwarding the talk about shared experiences": one party makes minimal reference to past shared experience (e.g., John says *Remember Mary's brother?*), and the other displays their recognition of it, takes it up and forwards it in the conversation (Fred responds *Oh God, he's so strange, what about when he...*), thereby demonstrating the common ground. (Lerner 1992; Mandelbaum 1987; Maynard and Zimmerman 1984; cf. Enfield 2003)

— "Using improprieties and taking up the other's improprieties by using additional improprieties and/or laughter": cussing and other obscenities; laughter in response to such improprieties; shared suspension of constraints usually suppressed by politeness (Jefferson 1974).

At least the first three of these cases are squarely concerned with the strategic manipulation of information—the incrementing, maintaining, or presupposing of common ground—with consequences for the relationship and for its maintenance.[7] These are important candidates for local, culturally variant practices for maintaining social membership in one or another level (the examples in [5] being all definitive of "closer" relationships). Whether these are universal is an empirical question. It requires close analysis of social interaction based on naturally occurring, informal conversation across cultures and in different types of social–cultural systems.

I now want to elaborate with further examples of social practices from specific cultural settings that show particular attention to the maintenance of social relationships at various levels. In line with the theme of the chapter, they concentrate on the management of, or presupposition of, common ground, with both informational payoffs and social-affiliational payoffs.

A first example, from Schegloff (2007), is a practice that arises in the cultural context of Anglo-American telephone calls (at least before the era of caller ID displays). It hinges on the presumption that people in close social relationships should be able to recognize each other by a minimal voice sample alone. Here is an example:

(6) 1   ((ring))
     2   Clara: Hello
     3   Agnes: Hi
     4   Clara: Oh hi, how are you Agnes

This typical case displays an exquisite minimality and efficiency, which puts on mutual display to the interlocutors the intimacy of their relationship, thanks to the mutual presumption of person recognition based on minimal information. In line 1, Clara hears the phone ring. When she picks up, in line 2, she does not identify herself by saying who she is. She gives a voice sample carried by the generic formula "hello." If the caller is socially close enough to the callee, he or she will recognize her by her voice (biased by expectation, given that one usually knows who one is calling). On hearing this, Schegloff explains, by supplying the minimal greeting response "Hi" in line 3, the caller "claims to have recognized the answerer as the person they meant to reach." (Otherwise–i.e., if the caller did not recognize the answerer–he or she would have to ask, or at least ask for confirmation; e.g., "Clara?") At the same time, the caller in line 3 is reversing the direction of this minimal-identifying mechanism, providing "a voice sample to the answerer from which callers, in effect, propose and require that the answerer recognize them." In this seamless and lightning-fast exchange, these interactants challenge each to recognize the other given the barest minimum of information, and through the course of the exchange each of them claim to have achieved that recognition. (Clara not only claims but demonstrates recognition by producing Agnes's name in line 4.) Were they not to recognize who was calling on the basis of a small sample of speech like "hi"—which, after all, was produced on the presumption that the quality of the voice should be sufficient for a close social associate to identify the person—they would pay a social price of disaffiliation via a betrayal of distance and lack of intimacy ("What? You don't recognize me?!;" cf. Schegloff 2007).

Consider a second example, another practice by which social interactants identify persons. In English, when referring to a nonpresent person in an informal conversation, a speaker may choose whether to use bare first name (John) as opposed to some fuller name (John Smith) or description (my attorney, Bill's brother, that guy there; Sacks and Schegloff [1979]2007, Enfield and Stivers 2007). The choice depends on whether it is in speaker and addressee's common ground who "John" is and whether he is openly known to this speaker–addressee pair as John. The choices we make will, in general, reflect the level of intimacy and intensity of social relations among speaker, addressee, and referent, and this more directly concerns the common ground of speaker and addressee. In my example (Fig. 7), Kou (left) has just arrived at his village home, having been driven from the city (30 or so km away) in a pickup. He has brought with him a load of passengers, mostly children, who have now scattered and are playing in the grounds of his compound. Saj (right), a neighbor of Kou, has just arrived on the scene.

Figure 7.

Saj asks Kou how many people were in the group that has just arrived with Kou's vehicle, following this up immediately by offering a candidate set of people: "Duang's lot" (line 1). The named referent—Duang—is Kou's third daughter.[8] Kou responds with a list of those who have arrived with him, beginning by listing four of his own daughters by name (lines 2–3), then mentioning two further children (line 4):

(7)
1 S maa2   cak2   khon2 niø —   sum1   qii1-duang3
     kaø   maa2
     come   how_many   person TPC   group   F.NONRESP-D
     T.LNK   come
     How many people have come?—Duang's lot have come?
2 K qii1-duang3 —   qii1-daa3,   qii1-phòòn2
     F.NONRESP-D   F.NONRESP-D   F.NONRESP-P
     Duang—Daa, Phòòn.

3   maa2   bet2   lèq5,   qii1-khòòn2van3
     come   all   PRF   F.NONRESP-K
     All have come, Khòònvan.

4   dêk2-nòòj4   maa2   tèè1 paak5san2   phunø   qiik5   sòòng3
     khon2
     child-small   come   from P   DEM.FAR more   two
     person
     Kids from Paksan, another two.

It is in the common ground that Kou's own four children are known to both
Kou and Saj by their first names. Kou is therefore able to use the four
children's personal names in lines 2–3 to achieve recognition. In line 4,
Kou continues his list, with two further children who have arrived with
him. These two are not his own, are not from this village, and are presumed
not to be known by name to Saj. They are children of Kou's brother and
sister, respectively, who both live in Kou's mother's village Paksan, some
distance away. Kou refers to them as "kids from Paksan." The reason he
does not he refer to these two children by name is that he figures his
addressee will not recognize them by name—their names, as ways of
uniquely referring to them, are not in the common ground. But although Saj
certainly will not recognize the children by name, he will recognize their
village of origin by name (and further, will recognize that village to be
Kou's village of origin, and the home of Kou's siblings). So Kou's solution
to the problem of formulating reference to these two children—in line 4—is
to tie them to one sure piece of common ground: the name of the village
where a host of Kou's relatives are (openly, mutually) known to live.

However, it appears that Kou's solution in line 4 is taken—by Saj—to
suppose too little common ground. Although Saj would not know the
names of these Paksan children, he does know the names of some of Kou's
siblings from Paksan. This is common knowledge, which could form the
basis of a finer characterization of these children's identities than that
offered in line 4. What immediately follows Kou's vague reference to the
two children by place of origin in line 4 is Saj's candidate offer of a more
specific reference to the children. Saj's candidate reformulation (line 5 in
[8], below) links the children explicitly to one of Kou's siblings, referring
to him by name. This guess, which turns out to be not entirely correct,
succeeds in eliciting from Kou a finer characterization of the children's
identities (line 6). This new characterization presupposes greater common
ground than Kou's first attempt did in line 4, yet it remains a step away in

implied social proximity from that implied by Kou's first-name formulations to his own children in lines 2–3, above:

(8)   (Follows directly from (7))
5   S   luuk4   qajø-saaj3
      child   eBr-S
      Children of Saaj?
6   K   luuk4   bak2-saaj3phuu5 nùng1,   luuk4–qii1-vaat4sanaa3
      phuu5 nùng1
      child   F.NONRESP-S   person   one   child F.NONRESP- V
      person   one
      Child of Saaj, one, child of–Vatsana, one.

The contrasts between the three ways to formulate reference to a person— by first name in lines 1–3, via place of origin in line 4, via parent's name in lines 5–6 —represent appeals to common ground of different kinds, and different degrees. They are indicative of, and constitutive of, different levels of social familiarity and proximity. This example shows how such expression of these levels of familiarity can be explicitly negotiated within the very business of social interaction. Kou's reference to the two children from Paksan in line 4 was constructed differently to the references to his own children in lines 2–3, but Saj effectively requested, and elicited, a revision of the first-attempt formulation in line 5, thereby securing a display of greater common ground than had a moment before been presupposed.[9]

A third example involves two men in a somewhat more distant relationship. This is from an exchange between the two men pictured on the left of Figs. 2–4, above. (I call them Foreground Man [FM] and Background Man [BM].) The men hardly know each other, but are of a similar age. The younger sister of BM's younger brother's wife is married to the son of FM. The two men seldom meet. Their kinship ties are distant. Their home territories—the areas about which they should naturally be expected to have good knowledge—overlap partially. They originate in villages that are a day's travel apart. This is far enough to make it likely that they have spent little time in each other's territory, but it is not so far that they would be expected to not have ever done so. The common ground at stake, then, concerns knowledge of the land.

The conversation takes place in the village of FM. This is therefore an occasion in which BM is gathering firsthand experience beyond his home territory. It may be inferred from the segment we are about to examine that

FM wants to display his familiarity with BM's territory. The point of interest in this conversation is a series of references to a geographical location close to BM's home village, but which FM apparently knows well about. Although the men are discussing medicinal herbs, BM mentions an area in which certain herbs can be found. His first mention of the place is by name: Vang Phêêng.[10] As with reference to persons (see previous example), the use of the bare name in first mention presupposes recognizability or identifiability (Schegloff 1972). This identifiability is immediately confirmed by FM's reply of "Yeah, there's no shortage (of that herb) there." There is then over a minute's further discussion of the medicine, before the following sequence begins:[11]

(9)
```
1 BM haak4   phang2-khii5    kaø   bòø qùt2   juu1    [thèèw3-
     root     P-K                   T.LNK NEG lacking at      area
     Hak phang khii (a type of medicinal root) is plentiful, at the area of-
2 FM                                         [qee5
                                             yeah
                                             Yeah,
3     kaø   cang1 vaa1 faaj3   vang2-phêêng2  faaj3   ñang3   qooj4
      T.LNK  so say  weir      VP             weir    INDEF   INTJ
      Like I said, Vang Phêêng Weir, whatever weir, oh.
4 BM m5
     mm
     Mm.
5 FM bòø     qùt2    lèq5,   faaj3   qanø-nan4       naø
     NEG     lacking PRF     weir    MC.INAN-DEM     PCL.PERIPH
     It's not lacking (medicinal roots and herbs), that weir.
6     tè-kii4 haak5 vang2 phêêng2 nanø        tèø-kii4 khaw3 paj3 tèq2-
      tòòng4
      before pcl   VP            TPC.NONPROX before  3PL.B go
      touch
      Before, Vang Phêêng, before for them to go and touch it
7     bòø daj4, paa1-dong3 man2        lèwø dêj2
      NEG can   forest     3.NONRESP   PRF  FAC.NEWS
      was impossible, the forest of it_{non-respect}, you know.
```

In line 1, BM mentions a type of herbal medicine, saying that it is plentiful. He is about to mention the location in which it is plentiful, as projected by the use of the locational marker glossed in line 1 as "at." Not only does FM anticipate this, but also anticipates *which* location it is that BM is about to mention (in a form of anticipation directly related to that in the more simple example shown above in Figs. 2–4), namely Vang Phêêng Weir (line 3). (cf. Lerner 1996 on collaborative turn completion.) This is confirmed by BM's acknowledgement marker "mm" in line 4. Again, we see a dance of display of common ground, by anticipation of what the current speaker is going to say. FM goes on to comment in lines 6–7 that in the old days it was impossible to collect medicinal herbs from the area.

The element of special interest here is the pronoun *man2* "it" in bold face in line 7. There is no local antecedent for this pronoun. The speaker is using a locally subsequent form in a locally initial position (Fox 1987; Schegloff 1996), with a subsequent major risk of not succeeding in getting recognition. How do his addressees know what he is talking about? (We get evidence that BM at least claims to follow him, as we see BM in the video doing an acknowledging "head toss"—something like a nod—directed to FM just as the latter utters line 7.) A couple of lines ensue (omitted here to save space), which finish with FM repeating that in the old days it was impossible to get medicinal herbs out of there. Then, FW contributes:

(10)
```
8 FW khuam2        phen1    haaj4 niø naø
        reason     3.P      angry TPC TPC.PERIPH
     Owing to it's_respect being angry?
9 FM qee5 — bòò1 mèèn2 lin5 lin5   dêj2,   phii3 vang2 phêêng2 niø
        yeah   NEG be   play play  FAC.NEWS spirit V      PCL
     Yeah—It's not playing around you know, the spirit of Vang Phêêng.
```

Line 8, uttered by FW (BM's wife) partly reveals her analysis of what FM is saying, and specifically of what he was referring to by the 3rd person singular pronoun *man2* in Line 7. She, too, uses a 3rd-person singular pronoun, but her choice is the polite *phen1*. She suggests that the previous difficulties in extracting herbs is because of "the anger of it." Someone who lacks the relevant cultural common ground will have no way of knowing that the referent of "it" is the spirit owner of Vang Phêêng. This is not made explicit until it seems obvious that everyone already knows what the speaker has been talking about—that is, as a demonstrative afterthought in line 9.

This exchange reveals to the analyst the extent to which recognition of quite specific references can be elicited using very minimal forms for reference when those involved in the social interaction share a good deal of common ground (cf. [3] and Figs. 5–6, above). It also makes important indications to the participants themselves. They display to each other, in a way hardly possible to fake, that they share specific common ground. In line 3 of (9), FM anticipates what BM is going to say, and says it for him. In line 7, FM uses a nearly contentless pronoun to refer to a new entity in the discourse, relying entirely on shared knowledge and expectation to achieve successful recognition.[12] In line 8, FW displays her successful recognition of the referent introduced by FM in line 7, by making explicit something about the referent that up to this point has been merely implied. By the economy and brevity of these exchanges, these individuals display to each other—and to us as onlookers—that they share a great deal of common knowledge, including common knowledge of the local area (and the local biographical commitment this indexes), and membership in the local culture. This may be of immense value for negotiating the vaguely defined level of interpersonal relationship pertaining between the two men, whose only reason for interacting is their affinal kinship. In conversing, they test for, and display common ground, and through the interplay of their contributions to the progressing trajectory of talk demonstrate a hard-to-fake ability to know what is being talked about before it is even mentioned.

8. Conclusion

This chapter has proposed that the practices by which we manage and exploit common ground in interaction demonstrate a personal commitment to particular relationships and particular communities, and a studied attention to the practical and strategic requirements of human sociality. I have argued that the manipulation of common ground serves both interactional efficacy and social affiliation. The logic can be summarized as follows. Common ground—knowledge openly shared by specified pairs, trios, and so forth—is by definition socially relational, and relationship defining. In an informational dimension, common ground guides the design of signals by particular speakers for particular recipients, as well as the proper interpretation by particular recipients of signals from particular speakers. Richer common ground means greater communicative economy, because it enables greater ampliative inferences on the basis of leaner coded signals. In a social-affiliational dimension, the resulting streamlined, elliptical interaction has a property that is recognized and exploited in the ground-level management of social relations: these indices of common ground are a means of publicly displaying, to interactants and onlookers alike, that the requisite common ground is indeed shared, and that the relationship constituted by that degree or kind of common ground is in evidence.

In sum, common ground is as much a social-affiliational resource as it is an informational one. In its home disciplines of linguistics and psychology, the defining properties of common ground concern its consequences in the realm of reference and discourse coherence. But sharedness, or not, of information, is essentially social. Why else would it be that if I were to get the promotion, I had better tell my wife as soon as I see her (or better, call her and let her be the first to know), whereas others can be told in due course (my snooker buddies), and yet others need never know (my dentist)? The critical point, axiomatic in research on talk in interaction yet alien to linguistics and cognitive science, is that there is no time out from the social consequences of communicative action.

Acknowledgements

This chapter was previously published in 2006 in the volume *Roots of human sociality* (Oxford: Berg , edited by NJ Enfield and SC Levinson), with the title 'Social Consequences of Common Ground'. The present version includes minor revisions. I thank Berg Press for granting permission to reproduce the chapter in this volume. I would like to acknowledge a special debt to Bill Hanks, Steve Levinson, Paul Kockelman, Tanya Stivers, Herb Clark, Chuck Goodwin, John Heritage, and Manny Schegloff, along with the rest of my colleagues in the Multimodal Interaction Project (MPI Nijmegen)—Penny Brown, Federico Rossano, JP de Ruiter, and Gunter Senft—for helping me develop my thinking on the topics raised here. I received helpful commentary on draft versions from Steve Levinson, Tanya Stivers, as well as Jack Sidnell and two other anonymous reviewers. None are responsible for errors and infelicities. I gratefully acknowledge the support of the Max Planck Society. I also thank Michel Lorrillard for providing me with a place to work at the Vientiane centre of l'École française d'Extrême-Orient, where final revisions to this chapter were made. Finally, I thank the entire cast of contributors to the Roots of Human Sociality symposium at the village of Duck, on North Carolina's Outer Banks, October 2004. Lao transcription and abbreviations follow Enfield (2007b).

Notes

1. See also Schiffer (1972), Sperber and Wilson (1995), D'Andrade (1987:113), Searle (1995:23–26), Schegloff (1996:459), Barr and Keysar (2004). Although analysts agree that humans can construct and consult common ground in interaction, there is considerable disagreement as to how pervasive it is (see discussion in Barr and Keysar 2004).
2. By *hypothesis,* I do not mean that we need consciously or explicitly entertain candidate accounts for questions like whether our colleagues will wear clothes to work tomorrow, or whether the sun will come up, or whether we will stop feeling thirsty after we have had a drink (saying "Aha, just as I suspected" when verified). But we nevertheless have models of how things are, which, most importantly, are always accessible, and become visible precisely when things go against our expectations (Whorf 1956). In order for this to work, we need some kind of stored representation, whether mental or otherwise embodied, which accounts for our expectations.
3. Steve Levinson points out the relevance of the great spatial distance between BW and the basket. Her reach has a long way to go when FW acts on the inference derived from observing her action. It may be that BW's stylized reach was overtly communicative, designed to induce recognition of intention, and the perlocutionary effect of causing FW to pass the basket (functioning, effectively, as a request).
4. The phrasing appropriates Slobin's thinking-for-speaking idea: that "language directs us to attend—while speaking—to the dimensions of experience that are enshrined in grammatical categories" (Slobin 1996: 71).
5. There is some controversy as to the extent to which we do audience design and assume its having been done. By a frugal cognition view, audience design is heavily minimized, but all analytical positions acknowledge that high-powered inference must at the very least be available when required (Barr and Keysar 2004; cf. Goodwin 2006, Hutchins 2006, Danziger 2006).
6. This is the corollary of the impossibility of pretending to possess common ground when you do not: witness the implausibility of fictional stories in which characters assume other characters' identities and impersonate them, living their lives without their kin and closest friends detecting that they are imposters (e.g., the reciprocal face transplant performed on arch enemies Castor Troy and Sean Archer in *Face/Off*, Paramount Pictures, 1997).
7. More work is needed to understand how the use of profanities works to display and constitute "close" social relations. Presumably, the mechanism is that "we can't talk like that with everybody." So, it is not a question of the symbolic content of the information being exchanged, but its register, its format. Compare this with more sophisticated ways of displaying social affiliation in the animal world, such as the synchronized swimming and diving that closely affiliated porpoises employ as a display of alliance

(Connor et al. 2000:104). It is not just <u>that</u> these individuals are swimming together, but, in addition, <u>how</u> they are doing it.

8. Like the others in this list of names, Duang is socially "lower" than both the participants, and accordingly, her name is prefixed with the female nonrespect prefix qii1-; cf. Enfield (2007a, b).

9. I gratefully acknowledge the contribution of Manny Schegloff and Tanya Stivers to my understanding of this example.

10. The Lao word *vang* refers to a river pool, a section of river in which the water is deep and not perceptibly flowing, usually with thick forest towering over it, producing a slightly spooky atmosphere, of the kind associated with spirit owners (i.e., ghosts or spirits that "own" a place, and must be appeased when traveling through; see Enfield 2008). The same place is also called Faaj Vang Phêêng (faaj means "weir"; the deep still water of Vang Phêêng is a weir reservoir).

11. Vertically aligned square brackets indicate overlap in speech.

12. This is comparable with the use of *him* in the opening words of Paul Bremer's highly anticipated announcement at a Baghdad news conference in December 2003 of the capture of Saddam Hussein: "Ladies and gentlemen, we got him."

References

Baron-Cohen, Simon
    1995.   *Mindblindness: An essay on autism and Theory of Mind.* Cambridge, MA: MIT Press.

Barr, Dale J., and Boaz Keysar
    2004   Making sense of how we make sense: The paradox of egocentrism in language use. In *Figurative language processing: Social and cultural influences.* Herb Colston and Albert Katz (eds.), 21–41. Mahwah, NJ: Erlbaum.

Buys, Christian J., and Kenneth L. Larson
    1979   Human sympathy groups. *Psychology Reports* 45:547–553.

Carruthers, Peter and Peter K. Smith (eds.).
    1996   *Theories of Theories of Mind.* Cambridge: Cambridge University Press.

Clark, Herbert H.
    1996   *Using language.* Cambridge: Cambridge University Press.

Cohen, Dov
    1999   Adding insult to injury: Practices of empathy in an infertility support group. Ph.D. diss. School of Communication, Rutgers University.

Connor, Richard C., Randall S. Wells, Janet Mann, and Andrew J. Read
    2000   The Bottlenose Dolphin: Social relationships in a fission-fusion society. In *Cetacean societies: Field studies of dolphins and whales.* Janet Mann, Richard C. Connor, Peter L. Tyack, and Hal Whitehead (eds.), 91–126. Chicago: Chicago University Press.

D'Andrade, Roy D.
    1987   A Folk Model of the Mind. In *Cultural models in language and thought.* Dorothy Holland and Naomi Quinn (eds.), 112–148. Cambridge: Cambridge University Press.

Drew, Paul, and Kathy Chilton
    2000   Calling just to keep in touch: Regular and habitualised telephone calls as an environment for small talk. In *Small Talk.* J. Coupland (ed.), 137–162. Harlow: Pearson Education.

Dunbar, Robin I. M.
    1993   Coevolution of neocortical size, group size, and language in humans. *Behavioral and Brain Sciences* 16:681–735.
    1996   *Grooming, gossip and the evolution of language.* London: Faber and Faber.
    1998   The social brain hypothesis. *Evolutionary Anthropology* 6:178–190.

Dunbar, Robin I. M., and M. Spoors
    1995   Social networks, support cliques, and kinship. *Human Nature* 6:273–290.

Enfield, Nicholas J.
    2002   Cultural logic and syntactic productivity: Associated posture constructions in Lao. In *Ethnosyntax: Explorations in culture and grammar.* Nicholas J. Enfield (ed.), 231–258. Oxford: Oxford University Press.
    2003   The definition of what-d'you-call-it: Semantics and pragmatics of recognitional deixis. *Journal of Pragmatics* 35:101–117.
    2005   The body as a cognitive artifact in kinship representations. Hand gesture diagrams by speakers of Lao. *Current Anthropology* 46(1):51–81.
    2006   Evidential interrogative particles in Lao. (Typescript) *Language and Cognition Group,* MPI Nijmegen, March 2006.
    2007a   Meanings of the unmarked: how 'default' person reference does more than just refer. In *Person reference in interaction.* Nicholas J. Enfield and Tanya Stivers (eds.), 97–120. Cambridge: Cambridge University Press.
    2007b   *A grammar of Lao.* Berlin: Mouton de Gruyter.
    2008   Linguistic categories and their utilities: the case of Lao landscape terms. *Language Sciences,* 30.2/3, 227–255.

Enfield, Nicholas J., and Tanya Stivers (eds.).
2007   *Person reference in interaction: Linguistic, cultural, and social perspectives.* Cambridge: Cambridge University Press.
Fox, Barbara A.
1987   *Discourse structure and anaphora: Written and conversational English.* Cambridge: Cambridge University Press.
Garfinkel, Harold, and Harvey Sacks
1970   On formal structures of practical actions. In *Theoretical sociology: Perspectives and developments.* John C. McKinney and Edward A. Tiryakian (eds.), 337–366. New York: Meredith.
Gigerenzer, Gerd, Peter M. Todd, and The ABC Research Group
1999   *Simple heuristics that make us smart.* Oxford: Oxford University Press.
Goffman, Erving
1967   *Interaction ritual.* New York: Anchor Books.
1971   *Relations in public.* New York: Harper & Row.
1981   Reprint. *Forms of talk.* Philadelphia: University of Pennsylvania Press. 1976.
Goodwin, Charles
1994   Professional vision. *American Anthropologist* 96(3):606–633.
1996   Transparent vision. In *Interaction and grammar.* Elinor Ochs, Emanuel A. Schegloff, and Sandra A. Thompson (eds.), 370–404. Cambridge: Cambridge University Press.
2000   Action and embodiment within situated human interaction. *Journal of Pragmatics* 32:1489–1522.
Goody, Esther N.(ed.).
1995   *Social intelligence and interaction: Expressions and implications of the social bias in human intelligence.* Cambridge: Cambridge University Press.
Grice, H. Paul
1975   Logic and conversation. In *Speech acts.* Peter Cole and Jerry L. Morgan (eds.). 41–58. New York: Academic Press.
1989   *Studies in the way of words.* Cambridge, MA: Harvard University Press.
Heritage, John, and J. Maxwell Atkinson
1984   Introduction. In *Structures of social action: Studies in conversation analysis.* J. Maxwell Atkinson and John Heritage (eds.), 1–15. Cambridge: Cambridge University Press.
Hill, R. A., and Robin I. M. Dunbar
2003   Social network size in humans. *Human Nature* 14:53–72.
Hutchins, Edwin
1995   *Cognition in the wild.* Cambridge, MA: MIT Press.

Hutchins, Edwin, and Brian Hazlehurst
   1995   How to invent a shared lexicon: The emergence of shared form-
           meaning mappings in interaction. In *Social intelligence and
           interaction: Expressions and implications of the social bias in
           human intelligence.* Esther Goody (ed.), 53–67. Cambridge:
           Cambridge University Press.
Hutchins, Edwin, and Leysia Palen
   1993   Constructing meaning from space, gesture, and speech. In
           Discourse, tools, and reasoning: Essays on situated cognition.
           Lauren B. Resnick, Roger Säljö, Clotilde Pontecorvo, and Barbara
           Burge (eds.), 23–40. Berlin: Springer-Verlag.
Jefferson, Gail
   1974   Error Correction as an Interactional Resource. *Language in Society*
           2:181–199.
   1978   Sequential aspects of storytelling in conversation. In *Studies in the
           organization of conversational interaction.* J. Schenkein(ed.), 219–
           248. New York: Academic Press.
Jefferson, Gail, and John R. E. Lee
   1980   *End of Grant Report to the British Ssrc on the analysis of
           conversations in which "troubles" and "anxieties" are expressed.*
           Ref. Hr 4802. Manchester: University of Manchester.
Kita, Sotaro. (ed.)
   2003   *Pointing: Where language, cognition, and culture meet.* Mahwah,
           NJ: Erlbaum.
Kockelman, Paul
   2005   The semiotic stance. *Semiotica* 157:233–304.
Lerner, Gene H.
   1992   Assisted storytelling: Deploying shared knowledge as a practical
           matter. *Qualitative Sociology* 15:24–77.
   1996   On the "semi-permeable" character of grammatical units in
           conversation: Conditional entry into the turn space of another
           speaker. In *Interaction and grammar.* Elinor Ochs, Emanuel A.
           Schegloff, and Sandra A. Thompson (eds.), 238–276. Cambridge:
           Cambridge University Press.
Levinson, Stephen C.
   1995   Interactional biases in human thinking. In *Social intelligence and
           interaction: Expressions and implications of the social bias in
           human intelligence.* Esther Goody (ed.), 221–260. Cambridge:
           Cambridge University Press.
   1997   From outer to inner space: Linguistic categories and non-linguistic
           thinking. In *Language and conceptualization.* Jan Nuyts and Eric
           Pederson (eds.), 13–45. Cambridge: Cambridge University Press.

2000   *Presumptive meanings: The theory of generalized conversational implicature.* Cambridge, MA: MIT Press.

Lewis, David K.
1969   *Convention: A philosophical study.* Cambridge, MA: Harvard University Press.

Mandelbaum, Jennifer
1987   Couples sharing stories. *Communication Quarterly* 352:144–170.

Maynard, Douglas W., and Don Zimmerman
1984   Topical talk, ritual, and the social organization of relationships. *Social Psychology Quarterly* 47:301–316.

Mead, George H.
1934   *Mind, Self, and Society from the Standpoint of a Social Behaviorist.* Charles W. Morris (ed.), Chicago: University of Chicago Press.

Miller, Geroge A.
1951   *Language and communication.* New York: McGraw-Hill.

Molder, Hedwig te, and Jonathan Potter (eds.)
2005   *Conversation and Cognition.* Cambridge: Cambridge University Press.

Moore, Chris, and Philip Dunham (eds.)
1995   *Joint attention: Its origins and role in development.* Hillsdale, NJ: Erlbaum.

Morrison, J.
1997   Enacting involvement: Some conversational practices for being in relationship. Ph.D. diss. School of Communications, Temple University.

Nettle, Daniel
1999   Language variation and the evolution of societies. In *The evolution of culture: An interdisciplinary view.* Robin I. M. Dunbar, Chris Knight, and Camilla Power (eds.), 214–227. New Brunswick, NJ: Rutgers University Press.

Nettle, Daniel, and Robin I. M. Dunbar
1997   Social markers and the evolution of reciprocal exchange. *Current Anthropology* 38(1):93–99.

Norman, Donald A.
1991   Cognitive artifacts. In *Designing interaction: Psychology at the human-computer interface.* John M. Carroll (ed.), 17–38. Cambridge: Cambridge University Press.

Peirce, Charles S.
1965   Reprint *Collected papers of Charles Sanders Peirce, vol. 2: Elements of Logic.* Charles Hartshorne and Paul Weiss (eds.), Cambridge, MA: Belknap Press of Harvard University Press. 1932.

Pomerantz, Anita, and Jennifer Mandelbaum
  2005   Conversation analytic approaches to the relevance and uses of
          relationship categories in interaction. In *Handbook of language and
          social interaction,* K. L. Fitch and R. E. Sanders (eds.), 149–171.
          Mahwah, NJ: Erlbaum.
Rogoff, Barbara
  1994   *Apprenticeship in thinking: Cognitive development in social context.*
          New York: Oxford University Press.
Sacks, Harvey
  1974   An analysis of the course of a joke's telling in conversation. In
          Explorations in the ethnography of speaking. Richard Bauman and
          Joel Sherzer (eds.), 337–353. Cambridge: Cambridge University
          Press.
  1992   *Lectures on conversation.* London: Blackwell.
Sacks, Harvey, and Emanuel A. Schegloff
  1979   Two preferences in the organization of reference to persons in
          conversation and their interaction. In *Everyday language: Studies in
          ethnomethodology.* George Psathas (ed.), 15–21. New York:
          Irvington.
Schank, Roger C., and Robert P. Abelson
  1977   *Scripts, plans, goals, and understanding: An inquiry into human
          knowledge structures.* Hillsdale, NJ: Erlbaum.
Schegloff, Emanuel A.
  1972   Notes on a conversational practice: Formulating place. In *Studies in
          social interaction.* D. Sudnow(ed.), 75–119. New York: The Free
          Press.
  1982   Discourse as an interactional achievement: Some uses of "Uh Huh"
          and other things that come between sentences. In *Georgetown
          University Roundtable on Languages and Linguistics 1981;
          Analyzing Discourse: Text and Talk,* Deborah Tannen (ed.), 71–93.
          Washington, DC: Georgetown University Press.
  1992   Repair after next turn: The last structurally provided defense of
          intersubjectivity in conversation. *American Journal of Sociology*
          97(5):1295–1345.
  1996   Some practices for referring to persons in talk-in-interaction: A
          partial sketch of a systematics. In *Studies in anaphora.* B. Fox (ed.),
          437–485. Amsterdam: John Benjamins.
  1997   Third turn repair. In *Towards a social science of language, vol. 2:
          Social interaction and discourse structures.* G. R. Guy, C. Feagin, D.
          Schiffrin, and J. Baugh(eds.), 31–40. Amsterdam: John Benjamins.
  2002   Tutorial on membership categorization devices. To appear in *Issues
          in Conversation Analysis.* M. F. Nielsen and J. Wagner (eds.).

2007   Conveying who you are: The presentation of self, strictly speaking. In *Person reference in interaction.* N. J. Enfield and T. Stivers (eds.).

Schelling, Thomas C.
1960   *The strategy of conflict.* Cambridge, MA: Harvard University Press.
1978   *Micromotives and macrobehavior.* New York: W. W. Norton.

Schiffer, Stephen R.
1972   *Meaning.* Oxford: Clarendon Press.

Schutz, Alfred
1970   *On phenomenology and social relations.* Chicago: University of Chicago Press.

Searle, John R.
1995   *The construction of social reality.* New York: Free Press.

Slobin, Dan
1996   From "thought and language" to "thinking for speaking." In *Rethinking linguistic relativity.* John J. Gumperz and Stephen C. Levinson (eds.), 70–96. Cambridge: Cambridge University Press.

Smith, Neil V. (ed.)
1980   *Mutual knowledge.* London: Academic Press.

Sperber, Dan, and Diedre Wilson
1995   *Relevance: Communication and cognition, 2nd edition.* Oxford: Blackwell.

Sugawara, Kazuyoshi
1984   Spatial proximity and bodily contact among the Central Kalahari San. *African Study Monograph (supp.)* 3:1–43.

Tomasello, Michael
1999   *The cultural origins of human cognition.* Cambridge, MA: Harvard University Press.

Vygotsky, Lev S.
1962   Reprint. *Thought and language.* Cambridge, MA: MIT Press. 1934.

Whorf, Benjamin L.
1956   *Language, thought, and reality.* Cambridge, MA: MIT Press.

"Impeach or exorcise?" Or, what's in the (common) round?

Jacob L. Mey

Motto: "If you can't impeach him, exorcise him"

(from a message sent to Rep. Robert Drinan

(D-Mass.) by one of his constituents).

Source: TRB, *The New Republic*, February 9, 1974.[1]

1. Introduction

Linguists, sociologists, anthropologists, psychologists and all sorts of professional and lay people have been fascinated by the notion of *common ground*. Following Erving Goffman's original concept of *footing* as the place where one has one's feet implanted in common territory, common ground has become understood as the social situation of mutual awareness, with all the rights and duties inherent therein. The idea that there has to be a common ground (not just etymologically derived) for those who want to communicate, has been varied in innumerable ways, both to confirm its existence (Hanks, the Goodwins) and to criticize its absence (Rancière). One of the most fertile developments of the notion is due to the late Japanese linguist Akio Kamio, who in a series of publications has defended the idea that not everybody can say whatever he or she likes, regardless of proper footing. Similarly, John Heritage has advocated a stronger and more restricted use of the notion of *empathy*, to wit, a use that takes into account the limits of what or whom one can empathize with. Especially when it comes to verbal expressions of empathetic and other communicative activity, the familiar concept of *speech act* has shown itself to be insufficient to capture the variety of uses that are allowed or desirable in a particular situation. Concepts such as *pragmatic act* (Mey: 2001) and *situation bound utterance* (Kecskés: 2000, 2002) have been introduced recently to bridge the gap between what is said (duly or dutifully) and what is done (effectively or legally). The somewhat obsolete and currently less commonly employed speech act "to exorcise" provides an unusual instance of the old adage "saying it don't make it so" thus being given a new lease on life, but perhaps an unexpected one.

2. The notion of common ground

Common ground, in everyday speech, is usually taken to have to do with what my interlocutors and I have in common when it comes to our cultural, linguistic, and other backgrounds. More often than not, the background is defined in terms of *shared knowledge*, that is, the knowledge about the world that each of us brings to the conversation. Such shared knowledge is in our days heavily dependent on the media and their interpretation of the news. Thus, in order to be a valuable addition to the conversational situation called the morning coffee break, one has to be aware of what has been 'in the news' the night before. The person who has no such knowledge (perhaps due to the fact that he or she does not own a TV set or never watches the news), is by the same token excluded from the conversation and can be said to lack common ground with his or her conversational partners.

However, restricting common ground to what one knows, or has heard in, or remembered from, earlier conversations is certainly not the whole story. Shared knowledge was often thought of (e.g. by philosophers such as Stalnaker or anthropologists such as Levinson) to be based on shared presuppositions, that is, beliefs about the world that are *common* to the speaker and hearer.

These beliefs can be decisive when it comes to determining whether or not what is shared, is seen as true or false, whether it rhymes with one's understanding of the world, or contradicts it, and so on. But already Stalnaker noted that, as soon as we transcend the narrow frame of conversation, viewed as pure exchange of information, difficulties arise: "the difficulties come with contexts in which other interests besides communication are being served by the conversation" (Stalnaker 1974: 201) And those *other interests* are certainly not limited to the *truth conditions* of the utterances produced in conversation.

The French sociologist Jacques Rancière has in a number of works (e.g., 1995) tried to put to rest the specter of shared knowledge as the ultimate criterion for understanding. Rancière does this by appealing to an analogy with other realms of human activity: the military, the theatrical, the music, etc. His suggestion is to think of all these activities as happening on a *common scene*, understood as a *battlefield of ideas*. As I say in my 2001 book *Pragmatics*, commenting on Rancière's work:

> "The common scene is not simply a matter of agreeing on a common ground, or establishing some common definitions or some common conceptual framework. Rather, we are dealing with a contest here, a battle: while trying to establish their common ground, people incessantly engage in fights about issues thought to be 'common' but in reality originating in various kinds of misunderstandings." (Mey 2001: 212)

While for Rancière, the notion of common ground itself seems to be a misnomer, I would like to keep the concept, but extend it to not only comprise what is usually called *presuppositions,* but to also include the metapragmatic conditions for having those presuppositions, or common beliefs. Such conditions are called *metapragmatic* because they transcend the pragmatic domain where issues, beliefs, attitudes etc. are being held and put to work by the users of language. Common ground, in this sense, is thus typically a meta-concept, needed to explain why people in their conversational interaction sometimes fail to even *see* what their interlocutors are on about, precisely because there is something missing in their mental, moral, linguistic, psychological, social makeup that is present in the others'.

Put differently, what is needed is that the interlocutors (or interactants, more generally) are aware of each other's presuppositions, taken as the entirety of their mental etc. makeup; this is precisely what Thomas Scheff (2006, 2007) has identified as one of the essential features of Goffman's understanding of the notion of common ground: to wit, mutual awareness. The next section will discuss some of the implications of this awareness for the practice of language.

3. Mutual awareness and territory of information

The late Akio Kamio's ground-breaking work in pragmatics is centered around his theory of the "territory of information," as originally formulated in his Japanese Ph. D. dissertation (1987, 1990) and subsequent articles and books in English (1994, 1995, 1997). In order to explain why certain locutions are pragmatically impossible in Japanese under certain conditions, Kamio established two psychological scales, linearly displaying the location (or closeness) of speaker and hearer information, respectively.

Corresponding to these scales and their local representations are "two conceptual categories, called the speaker's and the hearer's territories of information" (Kamio 1994: 82).

Kamio's metaphor of closeness is intended to express the proprietary character of one's information. Close to the speaker, respectively the hearer, is information obtained directly through seeing or hearing an event, or information that pertains to the speaker or hearer's internal or external state, including beliefs, expertise, and knowledge about persons or events nearby. In such cases, a Japanese speaker will use a direct form in order to indicate that what is said is within the speaker's territory of information; in other cases, the speaker would have to express him- or herself indirectly, by including the information in question in a kind of parenthetical construction: "I hear that …," "it seems that …"

To use Kamio's example, a husband can state about his wife that she is 46 by saying:

Kanai wa 46 desu 'My wife is 46 years old'.

But it would be very strange to hear a husband say:

Kanai wa 46 desu-tte

or:

Kanai wa 46 da-soo desu. (Literally, 'I hear/supposedly my wife is 46 years old') (Kamio 1994: 75)

Here, the hearsay marker–*tte* or a quotative such as *soo* serves to indicate that the information conveyed is not in the speaker's territory of information.[2] Thus, in Japanese one cannot state anything directly about another person's private state of mind or body. Rather than saying to a person: "You are sad," one has to limit oneself to expressing one's opinion of the other's state, e.g. by saying: "To me, it looks like you're being sad." Such expressions of direct intervention, even if they are grammatically and semantically correct and seem to obey all of the rules, cannot be properly produced, because they *pragmatically* impinge upon another person's territory of information, that is to say, on information that the addressee controls.

In other words, Japanese speakers can only use the *direct* forms when expressing something that one has the right to say (speaking very generally), or which is in one's own territory of information (to use Kamio's terminology). Japanese hearers likewise control the choice of forms to be used when they are being addressed: direct vs. indirect.

Akio Kamio's theory (as expressed in his books) has a general validity; it not only concerns phenomena of Japanese and the rights involved in Japanese language use. His subsequent publications (in the *Journal of Pragmatics* and elsewhere) represent a successful attempt to introduce his way of pragmatic thinking into the study of other languages, such as English, as well. Kamio's theory fits in well with my suggestions about the right to use language, both in general and when special conditions prevail, and on how to deal with the necessity of defining and codifying the corresponding acts.

The next section will deal with a more general approach to the problems of speech acting, seen as exercises of rights. I will do this by first introducing my own theory of *pragmatic acts*, and then apply this to a particular case of pragmatic acting, one that is subject to rather special common ground conditions: the act of *exorcising*.

4.   Speech acts and pragmatic acts: Exorcising

Speech acts, as conceived by Austin, Searle, and Grice, occur in a vacuum of abstract propositions whose *value* (originally conceived of in terms of truth) resides in their point or intention, and in the effect they can be assigned when executed. To capture these abstract acts and to legislate their potential use, Austin, Searle and their followers developed a battery of conditions and criteria, determining the felicitous outcome of a particular act. Common to all these conceptualizations was the complete disregard for the users, the live utterers and receivers of the acts. In addition, when the user was mentioned at all, only speakers were taken into account; the hearers were conspicuously absent from most of traditional speech act theorizing (a trait also noticed by critics of speech act theory such as Sperber and Wilson 1995).

To remedy the deficits of the classical theory, various efforts have been undertaken. In the second edition of my book *Pragmatics: An Introduction* (2001), I have developed the wider notion of pragmatic act to replace the limited concept of speech act. Pragmatic acts are not restricted to speech, but include all the other elements of human communication: gestures, facial expressions, extralinguistic components, and so on. Mainly, they distinguish themselves from speech acts by being developed in a *situation of use*.

The situation that makes a speech act possible has never been seriously considered by the theorists except in a very remote way (compare the external felicity conditions on a speech act such as marrying (a couple), among others the proper signing and delivery of the marriage document by the parties before witnesses (Thomas 1996: 39). In contrast, I consider the situation as representing a *total user involvement*; by this I mean two things: one, the user involves the situation, by being an active (co-) participant and language user; two, and perhaps more importantly, the situation involves the user, by defining what he or she can do and say in the situation. Both aspects crucially depend on who defines the situation and what the empowerment of the participants is grounded in.

When Rep. Drinan was admonished by one of his constituents to exorcise President Nixon if he couldn't impeach him, the writer of the note probably did not reflect too much on what kind of speech act he or she was referring to. The time of writing, early 1974, was when everybody anxiously awaited the result of the investigation into the multiple frauds that had been committed and/or sanctioned under Nixon's supervision in order to cover up Watergate and related scandals. The specter of having to impeach a sitting president was raising its ugly head, with all the devastating consequences such an act might have on the political life of the nation and its relations to the outside world.

Besides, the process of impeaching a ruling president was unfamiliar to everybody involved; in US history, the last (and thus far the only, but unsuccessful) effort at presidential impeachment had happened a hundred and forty years earlier (Andrew Johnson, 1867–1868), so there were virtually no precedents. Yet for many the only way out of the Nixon morass was to remove the man himself forcefully but legally; but again, the obstacles seemed enormous. No wonder then that for some, the reasonable thing to do was to implore help from higher quarters, by having recourse to the old practice of driving out demons, known as exorcising. In addition, movies such as Polanski's *Rosemary's Baby* (1968) and Friedkin's *The Exorcist* (1973), fresh in the memory of millions of Americans, had, so to say, reintroduced the supernatural, and in particular the option of performing an exorcism, into the public consciousness.

Compared to the act of impeachment, with all its legal hurdles and political roadblocks, exorcising Nixon seemed to be an easier and more efficient way to get rid of him—that is, if one could get an exorcist to take on the job. Now, the function of exorcist is actually one of the four minor orders of the priesthood in the Catholic Church (the other three being that of *ostiarius* or [door guard], *catechista* or [catechist], and *acolythus* [acolyte][3]). By being ordained in this fractioned way, the future priests are given some part in the power of the office, even though the major orders (*subdiaconus* [subdeacon], *diaconus* [deacon], and *sacerdos* [priest]) and investiture with the full priestly functions are still some years away. In addition, none of these functions and powers can be exercised fully and legally without authorization by the competent higher authorities, the bishops, and eventually the Pope himself. Concretely, this means that although I personally was ordained an exorcist back in 1951, I have never performed an exorcism or even considered practicing one, knowing full well that I would never be given the authority to do so (which was exactly the young Jesuit, Fr. Damian's, problem in the movie *The Exorcist*).

As to pragmatic acts, they consist of two parts: an *activity part* and a *textual* one (Mey 2001: 222). In the activity part, one finds the speech acts and other, related acts (interactional, prosodic, psychological, physical, etc.), while the textual part contains the various features (tense, modality, deixis, etc.) that characterize the more or less linear sequence of linguistic units involved in the production of the pragmatic act.

When people are confronted with a pragmatic act, it is usually the *verbal* part of the act that attracts the most attention, not only among the general public, but also among the speech act analysts. In our society, the words one utters can be made the object of a domestic quarrel, a court litigation, and even a criminal prosecution. In the judicial procedures, it is of the utmost importance to establish what the accused or the witnesses actually said; what they *did* (e.g. by moving their eyebrows or smiling) is much harder to deal with in official terminology and jurisprudence.

In the particular case of the exorcism, what we find in the speech portion of the act are expressions such as "We adjure you, cursed dragon, and you, diabolical legions, by the living God, by the true God, by the holy God, to stop deceiving human creatures and pouring out to them the poison of eternal damnation…"and"… tremble and flee when we invoke the Holy and terrible Name of Jesus, this Name which causes hell to tremble, …"or "In the Name and by the power of Our Lord Jesus Christ, may you be snatched away and driven from the Church of God, …" (quoted from "A Simple Exorcism for Priests or Laity Published by Order of His Holiness Pope Leo XIII." Source: Holy Catholic Church 2000).

Here, the verbal part of the exorcising act contains expressions such as *adjure, invoke, in the name of, by the power of*, and so on. These speech act verbs and nominal invocations embody the act of exorcising, whose very expression reflects its verbal origin, the Greek verb *eks-horkízein*, (to exorcise), that defines the exorcism etymologically; the out-oathing itself is based on the Greek word *hórkos* (oath).[4]

In the quote above, the exorcism is practiced by naming and invoking the otherwordly authorities, adjuring them to come and snatch away the evil spirits; by doing so, one places one's pragmatic act, so to speak, straight into the protective and active sphere of the invoked authority by uttering words that place the utterer and the addressee(s), the demon-possessed persons or localities, in that sphere. And even regular everyday oaths always contain an element of displacement from the secular order. The oath-taker places himself outside the normal order of things and enters into the sacred realm of the extremes: life and death, heaven and hell: *This and more may be done to me, if I do not ..., May the devil take me, So help me God, May God damn my soul, Come hell and high water*, and so on.

Let's now go back to the original case, the message to the late Rep. Robert Drinan, S.J. In spite of the attention given, and the strength commonly attributed, to the exorcising acts verbal portion, it should be clear that the invocations by themselves, when they do not comply with the proper conditions (that is, among other things, if they are spoken without authorization) will remain without effect. This is why the person writing to Rep. Drinan could not have been too serious about the proposal: he or she must have known that as a good Jesuit, faithful to his special Jesuit vow of obedience to the Pope, Fr. Bob would never commit a pious transgression of the kind we saw in *The Exorcist*, where the young priest, Fr. Damian, willingly undertakes an unauthorized, but somehow successful exorcism, only to be bailed out by his older colleague, the experienced Jesuit-exorcist (who then dies of a heart attack). Subsequently, Fr. Damian lets the Devil enter his body instead of the possessed girl's, and (in a curious replay of the scene recounted in Luke 8:33), hurls himself out the window, just as those unfortunate swine in the Bible threw themselves down the precipice and into Lake Kinnereth. A situation such as the one described in the movie would never arise in our case, simply because none of the proper conditions for the act were fulfilled.

Recall now what I said earlier about the situation as a total involvement of the participants. Even if one does not want to practice an exorcism, saying words of an exorcizing nature may have the effect of changing one's involvement, one's footing, vis-à-vis one's interlocutors. When I utter words which, if spoken in the right context under the right authorization, could be used as an exorcism, my addressee, upon realizing this, will consequently change his footing towards me. The situation is similar to uttering a racial slur or some other insulting comment, and then saying: "Sorry, I didn't mean to insult you," or even "I take that back." In real life, as in pragmatics, there is no taking back: "Things have been said which cannot be forgotten," as Bonnie Prince Charles said in R. L. Stevenson's novel *Kidnapped*. In the situational involvement, the involved cannot *ad libitum* detach themselves from their dialectic embrace.

Involvement in a situation means that one not only has certain obligations, but also specific rights. While the classical theorists (e.g., Grice) have a lot so say about what can be expected of rational conversationalists, and spell out their obligations in some detail, the matter of rights in speech acting has not been dealt with in the same fashion. Yet, an important aspect of involvement has precisely to do with rights. The next section will deal with the right to speak.

5. Who has the right to speak?

Classical speech act theory has very little to say about rights. The principles, maxims, and rules all deal with the obligations speakers have when using a particular formula (such as "I promise") in order to ensure that the corresponding speech act becomes "happy" or "felicitous." There are conditions of sincerity, brevity, manner; and they all spell out what one has to do when uttering a speech act. Nowhere is there a mention of what makes the speaker have the right, not just the ability or competence, to utter such an act.

As an example, take the speech act of congratulating. I remember how, in graduate school at Copenhagen University, one of my fellow students, a guy by the name of Ib, upon attending a lecture by our professor (the famous Louis Hjelmslev, who also happened to be my Ph.D. adviser) walked up to him after the speech (we were gathering in the kitchen of the Linguistics Department to have some coffee), and uttered the words: "Professor Hjelmslev, may I congratulate you on a splendid lecture." I don't know what was more amazing about this incident: the fact that Hjelmslev (who admittedly at the time already was declining) acknowledged the compliment with a visibly pleased "thank you;" or the fact that the lecture in question actually was one of the run-of-the-mill talks that I had heard Hjelmslev deliver a number of times; or (perhaps first of all) the fact that such a congratulatory remark by a student seemed highly inappropriate to me, being uttered by a person who I thought did not have the right to say what he said.

The keen sense of impropriety I perceived was the kind that one expresses in words as "my toes curled" or "my flesh crept." Even some fifty years later, I can recall this sensation, and experience the same sense of embarrassment as I did then, while waiting for my cup of coffee in the old Linguistics Department's haunts, located in the center of Copenhagen. But why was I embarrassed, and what was wrong with the speech act uttered by my commilito Ib?

Here, the notion of footing is helpful once again. If one's footing, in the sense of Goffman, depends on one's placement in the social situation, there clearly was a disproportion involved here. The student and the teacher were indeed participating in a common social situation, the lecture: Ib as listener, Hjelmslev as speaker; but they participated in totally different ways. Hjelmslev was the authoritative deliverer of a speech, intended to be part of a curriculum leading to a qualification on the highest level (the degree of Ph. D.); Ib and I were students, participating in the event as beneficiaries of the professor's words, not as evaluators or critics. Our duty was to listen, take notes, and speak only when spoken to; we depended on our professor to evaluate *us*, while *he* did not (at least not directly) depend on us for our judgment and critique.

What Ib did, by his infelicitous speech act, was transgressing the boundaries of his ground, his footing. Congratulatory remarks offered in a wrong kind of context backfire (one cannot congratulate a person on the passing away of a pet, for example). Neither can a student congratulate a professor on what the professor is duty bound to do: deliver a substantial (even brilliant) lecture. In Kamio's terminology, referred to earlier, Ib did not have access to the *territory of information* that the professor rightfully could consider as his exclusive property. The situation is exactly parallel to that described by Heritage and Sefi with regard to giving advice: "advice giving carries problematic implications about the knowledge or competence of the intended recipient" and, I add, of the advice giver him- or herself (Heritage and Sefi 1986:389).

6. Epistemic communities and the right to empathy (Heritage 2007)

Interactional activities need to be considered in the total context of their occurrence. One such context, that of the medical interview, has been examined in a recent work edited by Heritage and Maynard (2006). Among the contributions in that volume, I would like to highlight the one by Anssi Peräkylä, who discusses the well-known asymmetry in interactional control that occurs when doctors and patients try to establish a "common ground" for a diagnosis. In fact, doctors and patients enter into a kind of "dilemmatic" relationship, as Billig (1986) once called it: the dilemma is that on the one hand, the doctor is the authority, and the patient wants to recognize him or her as such (including taking his or her advice for granted); on the other hand, the patient objects to being treated as a mere case, a problem to be resolved without the patient being able to provide what he or she thinks is crucial input. As Peräkylä remarks, "These dilemmas cannot be resolved by the parties trying to subscribe exclusively to one or the other set of conflicting ideas, but instead by balancing them in their everyday practice." (2006: 215)

To resolve the dilemma, Peräkylä (2006: 219) suggests that we focus on the symmetric and dialogic aspects of the diagnostic conversation, where the doctor systematically orients to his or her medical accountability by presenting evidence for the diagnosis, and the patients react by their active, knowledgeable interventions reflecting their own picture of their medical condition, in an active response to the diagnostic statements by the doctor. However, he also cautions that at the basis of the communication, there is a power differential: even a patient-centered medical interview cannot escape the societal and medical constraints embodied in the final, authoritative judgment of the condition: the diagnosis.

Let me now revert to the notion of *territory of information* introduced earlier. John Heritage, in a lecture at the Symposium About Language and Society Austin (SALSA XV), held in Austin, Texas in April 2007, emphasized the need to ask "whose territory is accessed" in cases like the above (congratulating, giving advice, and so on) (Heritage 2007). It is important to note that Kamio's original idea of territory of information needs to be extended to comprise also such things as knowledge and experience; information then could be considered convenient shorthand for all kinds of interactional activities.

As Heritage pointed out in his Austin speech, there is a need to construct what he called "epistemic communities" in discourse. Such communities basically cover the shared territories of the participants; those territories are defined and circumscribed by the common interests and goals that the participants define in cooperation, and by which they consequently are defined themselves. The *social participation* that Goffman (as we saw earlier) highlights in his concept of footing comprises relations that are not just about managing information, but establish what Nick Enfield has called "affiliations" (2007), or the "social consequences of common ground"(2006).

By the same token, however, these social relations involve a third party; as I have expressed it elsewhere (Mey1985: 336), whenever people communicate, society is the "silent," but by no means "sleeping" partner. In other words, the *social* construction is subject to *societal* constraints, in particular as they become visible and have to be obeyed in institutional talk, such as the one Peräkylä is referring to. This implies a limitation of the interactants' rights, including the right to use speech, to speech-act; moreover it implies a duty of speaking, not only in institutionalized surroundings but also in conversation in general. Conversation is not just a frivolous pastime: it is a serious duty, as well as a right, to be exercised in the social situations in which we find ourselves. In Heritage's speech, he referred to the right to show empathy in situations where empathy is expected and allowed; contrariwise, there are situations where one does not have the right to be empathetic.

As examples, consider what happens in official encounters when participants become overly empathetic, sometimes with catastrophic consequences for the success of the interaction, as when one tries to buy an official's favors by offering him money to help him buy medicine for his wife—whose condition one empathizes with and perhaps honestly wants to do something about, but of course for the wrong reasons and without the right to empathize. Claudia Caffi has, in a thoughtful contribution, analyzed exactly such cases, and in particular refers to the strange behavior of Prince Myshkin in Dostoyevsky's *The Idiot* as a prime example of "cooperation going overboard:" quality, quantity, manner, sincerity and all the other maxims are simply practiced without restraint; the excess cooperation spells the end of cooperation itself (Caffi 2007). As a result, the pathos of empathy turns into over-geared enthusiasm and uncontrolled logorrhea (as in Myshkin's case)[5] or – which perhaps is even worse – into cheap conversational bathos.

This, again, means that the situations determine and define what we can say, and that the utterances we produce, in turn define the situations, in a continuous dialectical interplay. The next section will deal with this in more detail.

7. Pragmatic acts and Situation Bound Utterances (SBUs): Limits of access

I started out by referring to the missing concept of "speaker's rights and duties" in speech act theory. In a more general sense, the classical theory can be said to be deficient in more than just this one aspect; or better, the missing aspect is missing because the framework itself is too narrowly conceived. This is the background against which the notion of pragmatic act was developed (as outlined above).

Harking back to the idea of common ground as it was defined in an earlier section, I would say that pragmatic acting basically is "using language on common ground," involving the other participants of the situation as well as the material and other conditions determining the "ground." In Herb Clark's (1996) terminology, this common ground involves three things: joint action, communication (linguistic and non-linguistic), and human activity in a broad, general sense; exactly the three most important aspects of what I call the *situation of use*.

Borrowing an analogy first developed by Mikhail Bakhtin (1981: 272–273), one could also say that classical speech act theory defines the acts from the inside out: given a speaker and his or her words, what can these words do? In contrast, what the theory of pragmatic acts does is to turn the question on its head: given a situation and the social conditions of the speaker and hearers, what can be said and done? In other words, I take a centripetal, rather than a centrifugal, view of users and situations: the situation determines the acts that a user can realize, rather than the acts realizing the independent intentions of the user.

Ultimately, however, since center and periphery, speaker and hearer, are intimately bound up as users in a situation of use, their relationship is a dialectic one. The acts that a user realizes contribute to the establishing or confirming of the situation, just as the situation establishes and confirms the rights and duties of the users. This aspect of speech acting is particularly clear in institutional contexts, as we have seen in the preceding: a doctor is as dependent on his or her patient for the outcome of the diagnosis as the patient is on the doctor for the final treatment; this is what I earlier (in section 3) have called the *total user involvement* of the situation, seen as a common ground.

This way of looking at the situation also prepares the way for a novel interpretation of that old, familiar and a bit trashed concept of context. Rather than restricting it to what's around in the text (sometimes called the "cotext"), a dynamic view of context combines what Anita Fetzer has called the "top-down" and the "bottom-up" aspects. According to her, "... natural-language communication [is] a dialogical, cooperative and collaborative endeavor, in which local meaning is negotiated in context" (Fetzer 2001).

In such a dynamic approach, we can either study the *micro-processes* of negotiating meaning or the *macro-processes* of societal intervention, realized respectively as unilateral, speaker-originated speech acts or collective, speaker-hearer interactive and dialectic dialogue. In this way, we can also distinguish between individual presuppositions and conditions and collectively shared presuppositions (called "co-suppositions" by Fetzer) and give the concept of context a new interpretation, which comes very close to what I have called the "situation of use." where one's interactional engagement dialectically oscillates between "individual sense-making and collective coherence" (Fetzer 2001).

The traditional notion of context, as it is conceived of by most linguists, has also come under fire from other quarters: those of applied linguistics and intercultural studies. It is well known that the culture of a nation is reflected in the way they conceptualize the world and express their conceptual categories in words. Without subscribing to a radical Sapir-Whorfian way of thinking, one can safely assume that words are the mirror of a culture, and that words, when used in the proper situation, may mean more than they superficially seem to indicate. Moreover, just as situations may indicate proper wordage, the words themselves can also, given the right conditions, indicate, or even create, a situation.

The American linguist Bruce Fraser once composed a list of what he thought were expressions that characterized a situation uniquely (personal communication). Among his examples were such expressions as "The check is in the mail" (unpaid bills), "It has never been driven except by an old lady who didn't go over 40 miles an hour" (used car sales), and others. Conversely, certain expressions may obtain a radically different meaning when uttered in a situation where a literal meaning would be inappropriate. For instance, Istvan Kecskés mentions the New York-originated expression "Get out of here" in the sense of "Don't put me on"—a fairly recent development. Witness the testimony of a native New Englander who visited his birthplace after having spent many years in Brazil and got this reply after he had told his taxi driver that he had lost both his parents and grandparents all within one half year (John Schmitz, pers. comm.; 2006); the same expression in other situations would carry the literal meaning of "Go away" (Kecskés 2000:614).

Such expressions used to be called *routine formulae* (Coulmas 1981) or similar things; however, as Kecskés points out, there is a deeper side to the matter. Kecskés calls the expressions in question "situation-bound utterances" (SBU), because they are bound to the situation in which they are normally used, and vice versa, help create and maintain those very situations. While they have "lost their compositionality and are no longer transparent semantically" (Kecskés 2000), they become very manifest *pragmatically*: they are, in Kecskés' words, "pragmatic idioms." As such, they definitively serve many of the functions that normally are ascribed to speech acts (such as greetings, requests, invitations, and so on); however, it is quite difficult (not to say impossible) to determine their character on the basis of some intuitively coherent partition of speech acts (much the same as it is the case for the so-called "indirect speech acts").

In contrast, SBUs can be compared fruitfully to my notion of pragmatic acts, inasmuch as they contain exactly the same components, except perhaps for the fact that SBUs appear as frozen phrases. But then again, in this respect they are like many of the metaphors we use in our daily language, where the processing is not a matter of dissecting the words individually, but rather of grasping the "salience" of the expression (to use a term coined by the Israeli linguist Rachel Giora; see especially Giora 2003).

According to Kecskés, such frozen expressions are only active at the sentence level, and do not play any role at the level of discourse. Here, a more nuanced view would perhaps argue that, inasmuch as these utterances are "pragmatic, rather than lexical units" (Kecskés 2000: 610), their role must be a pragmatic one, that is to say, they are instrumental in creating the situation of discourse in which they are used. This kind of discourse creativity is of course different from sentence- or utterance creativity, and Kecskés is right in pointing out the differences in this respect. But in a wider perspective, routine formulae and SBUs are a kind of mini-pragmatic acts, and contribute to the building and maintaining of the discourse in exactly the same way as do pragmatic acts in general.

Finally, above, I talked about the right to speak as an essential component (often neglected) of speech acting. As illustrated by the case of SBUs, this omission becomes even more glaringly evident (as everybody knows), whenever the restrictions on the use of such formulae are, so to speak, built in into the formulae themselves, to the extent that the formulae become emblematic for the situations they connote. As Heritage (2007) has pointed out for the case of empathy, not everybody has the right to empathize with everybody; routine formulae denoting empathy are bound to situations where empathy is legitimate. Such a limited access to situations and their corresponding linguistic expressions can only be accounted for in a broader, social and communicative framework, where social participation crucially depends on one's placement in a social situation: once again, we are faced with the common ground (Goffman's footing or Kamio's territory of information).

8.   Conclusion: Saying it don't make it so, or...?

Several decades ago, while I still was more or less actively working in computational linguistics, trying to practice my theoretical knowledge on stubborn pieces of lexical and pictorial material, it often happened that one in the group had a brilliant idea how to solve a particular problem. He or she would expatiate on the idea and outline in great detail how it could be implemented on the computer, and lead to a breakthrough in our efforts. The more seasoned in our group would always have a rather diffident attitude to such brainstorming, and one of the adages that used to be invoked on those occasions was the old word of wisdom: "Saying it don't make it so" (with an implied second part: "only doing it will").

While the proverb (based as it is on experience) obviously has its merits, we tend to overlook an important aspect of the spoken word: *saying,* too, is a way of *doing.* In the parlance of speech act theory, the illocutionary *force* of the act cannot be dissociated from the *words* embodying that act and its illocutionary point. Many of the speech acts referred to in the classical repertoire rely exactly on this "force," which is neither mysterious nor otherworldly or angelic, but simply inherent in the social structures that the words refer to and help maintain. This is why promises are so different from culture to culture; this is also why beliefs and faiths in a real sense can shake the world and change the face of the earth. Recall the example of the exorcism referred to in an earlier section: Exorcising, seen as a type of speech act, is a verbal function; its exercise, if it has to be successful, is bound to the conditions that the Supreme Exorciser (whoever we may prefer to think that is) has stipulated for its use. If those conditions are not observed, disastrous consequences may follow, as we have seen illustrated in the movie *The Exorcist*, where loss of life and sanity were incurred by the hapless practitioners who thought they could sidestep the rules in the service of the Greater Good.

Right indexes power, at least in principle. In a more nuanced formulation, power is a necessary, but not always a sufficient condition for exercising one's rights: the power has to be exercised in appropriate circumstances or else it may backfire. Also, speaking with power puts one under the obligation to use that power to the benefit of the people spoken to, on the penalty of disrupting the common involvement and having one's interlocutors deny their cooperation. Thus, the power inherent in the doctor's medical qualifications must be exercised in tandem with the patient's willingness to deliver his or her account through anamnesis and symptom description; when this doesn't happen, or happens infelicitously, we have the situation described by Heath as an "asymmetrical distribution of knowledge and competence" (1992: 263), by which the efficiency and results of the medical interview itself are put in the balance.

We see here how our words, embodied in pragmatic acts, work in tandem with societal power in any given social situation. An appeal to an authority (medical, political, educational, or other) can only work if the situation allows for the exercise of that power, as expressed through the pragmatic acts that characterize the situation. A failed appeal is worse than no appeal, because it cements the asymmetry that is inherent in the situation from the very beginning. This is what often makes the medical interview such a fruitless exercise; it also may cause people to choose not to exercise their civic rights in a politically charged situation, because such an exercise only creates hassles and usually produces few positive results.

Harking back to the old adage "Saying it don't make it so:" If we want to have our pragmatic acts (and not just those of exorcising a demon or a bad president) to be successful, we should seriously look into ways of saying things that *do* make it so, and in the process, establish and confirm our common ground.

Notes

1. Robert Drinan (1924-2006; "Uncle Bob," as he was called in the family), was a Jesuit priest and member of the US House of Representatives until he was ordered out of politics by Pope John Paul II in 1986. The message to Rep. Drinan was sent at the height of the discussions about whether or not to impeach President Richard M. Nixon for his role in the Watergate scandals.
2. Analogous examples illustrating the case of the hearer can be found in Kamio's articles.
3. The function of *lector* "reader" is not an ordained one.
4. English similarly relies on invocations of an oath-like character, such as *to adjure*; compare the general English term for dealing with the supra-natural, *conjuring (up)* (e.g., a ghost), which is based on the Latin root found in the verb *iurare* [to swear]; compare also our term *jury* for a group of people impaneled in force of an oath.
5. In the context, Dostoevskij characterizes "the idiot," Prince Lev N. Myshkin, as having brought himself in a state of "overly happiness" (*"rasscastlivilsja"*– the scare quotes are the author's own). And, following the episode where Myshkin inadvertently breaks a precious Chinese vase during one of his ranting monologues (*The Idiot* Bk. IV ch. vii), rather than letting himself be subdued by the tactful understanding of the other guests, he resumes his duty-bound conversational "cooperation," disregarding the various hints that are thrown his way, to end up in an epileptic paroxysm (1960: 605).

References

Bakhtin, Mikhail M.
   1981   Discourse in the novel. In *The Dialogic Imagination. Four Essays by M. M. Bakhtin*, Michael Holquist (ed.), 259-422. Austin: University of Texas Press.
Billig, Michael, Susan Condor, Derek Edwards, Mike Gane, David Middleton, and Alan R. Radley
   1988   *Ideological Dilemmas*. London: Sage Publications.
Caffi, Claudia.
   2007   When mitigation is missing: Myshkin, Chauncey G., Hölderlin. (MS submitted for publication)
Clark, Herbert H.
   1996   *Using language*. Cambridge University Press, Cambridge.

Coulmas, Florian. (ed.)
1981   Conversational routine: Explorations in standardized communication
         situations and prepatterned speech. *Rasmus Rask Series in
         Pragmatic Linguistics* Vol. 2. The Hague: Mouton..

Enfield, Nick J.
2006   Social consequences of common ground. In *Roots of human
         sociality: Culture, cognition, and interaction*, Nick J. Enfield and
         Stephen C. Levinson (eds.), 399-430. Oxford: Berg.
2007   *Relationship thinking and human pragmatics.* Paper delivered at the
         Workshop on Emancipatory Pragmatics, Japan's Women's
         University, Tokyo, March 2007.

Dostoevskij, Fedor M.
1960   *Idiot.* (The Idiot). Gosudarstvennoe izdatel'stvo xudozestvennoj
         literatury, Moskva. 1868.

Fetzer, Anita
2001   Context in natural-language communication: Presupposed or co-
         supposed? In *Modeling and Using Context.* Lecture Notes in
         Computer Science 21(16). Berlin: Springer: 449-452.

Giora, Rachel
2003   *On our mind: Salience, context, and figurative language.* Oxford:
         Oxford University Press.

Heath, Christian
1996   Diagnosis in the general-practice consultation. In *Talk at Work:
         Interaction in Institutional Settings*, Paul Drew and John Heritage
         (eds.), 235-267. Cambridge: Cambridge University Press.

Heritage, John.
2007   *Territories of Knowledge, Territories of Experience: (Not so)
         Empathic Moments in Interaction.* Keynote speech at the XVth
         Symposium about Language and Society, Austin (SALSA) Austin,
         TX. April 14, 2007.

Heritage, John, and Sue Sefi
1992   Dilemmas of advice. In *Talk at work: Interaction in institutional
         settings*, Paul Drew and John Heritage (eds.), 359-417. Cambridge:
         Cambridge University Press.

Holy Catholic Church
2000   A simple exorcism rite.
         http://www.truecatholic.org/exorcismsimple.htm (accessed on
         August 27, 2006).

Kamio, Akio
1990   Reprint. *Joohoo no nawabari riron* [Theory of the territory of
         information]. Ph.D. diss. Tsukuba University, Tokyo: Taishukan.
         1987.

1994   The theory of territory of information: The case of Japanese. *Journal of Pragmatics* 21(1):67–100.

1995   Territory of information in English and Japanese and psychological utterances. *Journal of Pragmatics* 24 (3):235–264.

1997   Theory of territory of information. *Pragmatics and Beyond*. 48. Amsterdam/Philadelphia: John Benjamins.

Kecskés, Istvan

2000   A cognitive-pragmatic approach to situation-bound utterances. *Journal of Pragmatics* 32: 605-625.

2002   Situation-Bound Utterances in L1 and L2. Berlin/New York: Mouton de Gruyter.

Mey, Jacob L.

1985   *Whose language? A study in linguistic pragmatics.* Amsterdam/Philadelphia: John Benjamins

2001   Reprint. *Pragmatics: An introduction* 2d ed. Oxford: Blackwell. 1993.

Peräkylä, Anssi

2006   Communicating and responding to diagnosis. In *Communication in medical care. Interaction between primary care physicians and patients.* (Studies in Interactional Sociolinguistics 20.), John Heritage and Douglas W. Maynard (eds.), 214-247. Cambridge: Cambridge University Press.

Rancière, Jacques

1995   *La mésentente.* Paris: Galilée.

Scheff, Thomas J.

2006   *Goffman unbound: A new paradigm for social science.* New York: Paradigm Press.

2007   Goffman and Beyond. Unpublished paper accessed at http://www.soc.ucsb.edu/faculty/scheff/55.htm. August 24, 2007.

Schmitz, John R.

2006   On the notions 'Native'/'Nonnative': a dangerous dichotomy for World Englishes. *RASK: International Journal of Language and Communication.* (23): 3-25.

Sperber, Dan and Deirdre Wilson

1995   Reprint. Relevance: Communication and cognition. 2d ed. Oxford: Blackwell.1986.

Stalnaker, Robert A.

1973   Pragmatic presuppositions. In *Semantics and Philosophy,* Milton Munitz and Peter K. Ungerer (eds.), 197-213. New York: New York University Press.

Thomas, Jenny
   1995   Meaning in interaction: An introduction to pragmatics. London: Longman.

Egocentric processes in communication and miscommunication[1]

Boaz Keysar

1. Introduction

Newly married, my wife and I visited my family for Passover. We were browsing through an English language bookstore in downtown Jerusalem, when my wife pointed to a table that had a variety of Hagadas, the text used during the Seder (the traditional Passover meal), and said "So, the Seder is going to be all in Hebrew?" "Of course" I replied and proceeded to look around. She didn't talk to me for a couple of days. Eventually, I understood why. What she meant was "let's buy a Hagada in English," because it was clear to both of us that she didn't know Hebrew. I understood her question as a request for information. In fact, she thought that her intention to get the book in English was so obvious, that I must have understood it. Given that, my response was plainly rude. In this paper I argue that my wife and I are not alone, and that this miscommunication is rooted in the systematic way we process language. To explain our behavior, I will show that communication in general proceeds in a relatively egocentric manner, with addressees routinely interpreting what speakers say from their own perspective, and speakers disambiguating their utterances with little consideration to the mental states of their addressees. Speakers also tend to overestimate how effectively they communicate, believing that their message is understood more often than it really is. I will present findings from my laboratory and from the literature that suggest such systematic causes for miscommunication.

2. Communication and cooperation

Most people, most of the time, think that what they say is pretty clear. Ambiguity is not routinely noted when people normally communicate. In contrast, linguists and psychologists who study the use of language notice potential ambiguity everywhere. The newspaper is a goldmine for unintended meanings, as in this recent classified ad: "Bedroom furniture – Triple dresser with mirror, armoire, one night stand." But students of language also know that even if it said "one nightstand," the text cannot be devoid of ambiguity because *every* text can have more than one meaning. Even a simple statement such as "this chocolate is wonderful" is ambiguous because it could be a statement of fact, an offer, a request for more, and so on. Despite such ubiquitous ambiguity, there are two reasons why people may not be confused. They use context for disambiguation, and they assume that the writer or speaker is a cooperative agent (Grice 1975). With both powerful tools, language users take a linguistic system that has a huge potential to fail, and use it successfully.

The cooperative principle explains why communication succeeds. Language users presume that their communication partner is cooperative, and use this to extract a specific meaning that preserves this assumption. What the partner believes, thinks and knows is central to this process. For instance, cooperativeness requires a certain level of informativeness. A speaker is expected to be informative in the sense that she is not providing too little information or too much information. When a colleague asks where I live, and I do not wish to offend him, I do not say "in Chicago" even though it is perfectly true. We work together; he obviously knows I live in Chicago. In this sense, what I know about what my colleague knows, and what I assume about what he doesn't know, should be central to what I say.

Not only must others' mental states be central to communication, but there is a good reason to believe that people have a unique ability to make inferences about these mental states quickly and accurately. Sperber and Wilson (2002) argued precisely that. Because conversation is so quick, with rapid turn taking and facile inferences, they conclude that the human mind is designed to take into account the beliefs of the other effortlessly and automatically. This would suggest the existence of a mental module that is dedicated to the consideration of beliefs during language processing (Fodor 1985).

In this paper I challenge these assumptions. I argue that when people communicate they do not routinely take into account the mental states of others, as the standard theory assumes. People don't rely on the beliefs and knowledge of their addressees to design what they say, and addressees do not routinely consider what the speaker knows to interpret what they hear. Of course, sometimes they might. But such consideration of the mental state of the other is not done systematically. So I will argue that when people succeed in avoiding ambiguity, it is not necessarily because they are following the principle of cooperation.

Why would language users behave in such a strange way that defies "common sense?" Why would they not do as they "should" and take into account systematically the mental state of their communication partner? The reason is that our own perspective, knowledge and beliefs, have priority over anything else we know about others' perspective, knowledge and beliefs (Decety and Summerville 2003; Epley et al. 2004). Our own perspective, then, does not allow us to follow the cooperative principle's assumption. Taking the perspective of the other requires considerable attention and effort. This, in turn, can explain miscommunication. Misunderstanding, then, is not what occasionally happens when random elements interfere with communication; it is not only a product of noise in the system. It can be explained systematically as a product of how our mind works.

3. Understanding egocentrically

Young children know how to speak before they know how to reason well about other's beliefs. Only at around four to five years of age can children distinguish between what they know and what others know (Wellman, 1990; Wellman, Cross and Watson 2001). Before age four they behave as if their own beliefs are shared by others. Their reasoning about mental states is relatively egocentric. Their private knowledge overwhelms their thinking. The most compelling demonstration of this is the false belief task (Perner, 1991; Perner, Leekam and Wimmer 1987). The child hides a candy together with Sally and then Sally leaves the room. The child then moves the candy to a different hiding location. When Sally returns to the room, the child is asked where Sally will look for the candy. Young children think that Sally will look for the candy where it really is, in the new hiding place. Probably because they know where it is and this private knowledge overwhelms their reasoning. Around age four, children start to distinguish what they know from what others know, and they are more likely to think that Sally will look for the candy in the old hiding place, where she believes it is. This developmental trajectory seems universal, as it is typical not only of Western children but also in places with a very different culture such as China (Sabbagh, Xu, Carlson and Moses, and Lee 2006), and even in isolated, pre-literate cultures (Avis and Harris 1991).

3.1. From childhood to adulthood

Though it seems that children's thinking is transformed from egocentric to allocentric, we have shown that the basic egocentric tendency persists through adulthood. In an experiment where subjects followed instructions, we investigated whether their interpretations of the instructions were egocentric (Epley, Morewedge and Keysar 2004). The subject sat across the table from a "director," and the director told the subjects what objects to move around on the table. For instance, there were two trucks, a large one and a smaller one, both visible to the subject and the director, and the director said "Move the small truck." As with the hidden candy task, there was a third, even smaller, truck, which was visible only to the subject but not to the director. We made it painfully clear to the subjects that the director could not see the smallest truck, and that he will not ask them to move it. If they are not egocentric, then they should not think that the director intended them to move that truck.

We found that children tended to interpret "the small truck" quite egocentrically. Young children reached for the truck that only they could see, almost half of the time. We also discovered an interesting similarity between children and adults, as well as an interesting difference. We found that the initial process of interpretation is identical for children and adults. By tracking subjects' gaze, we could tell which object they are considering as the intended one. Adult subjects were just as quick as young children to initially look at the hidden truck. This initial process, then, confounds what the subject can see and what the director can see. To eventually interpret the instructions as intended, the subjects must then recover from their egocentric interpretation and find an object that can be seen by the director. Children were much less effective in this recovery than adults. Once they found an egocentric referent, they took much longer than adults to find the intended one. Even more, children were less able to recover from it altogether. Once they looked at the hidden object, they were more likely to make an error and reach for it (fifty-one percent) than adults (twenty-one percent).

What we discovered, then, is that even though children are eventually able to represent the beliefs of others, this ability does not guide their interpretation of others' actions. Even adults initially behave as if they confound the knowledge of the other and their own, but eventually use their understanding of beliefs to correct their interpretation. In this sense, adults are not allocentric in how they understand others, they are just more practiced in overcoming an inherent egocentric tendency. The same is true for the very ability to think about beliefs (Birch and Bloom, 2007). Adults fail the false belief task if the task is a bit more complex. Five year olds are already able to predict that Sally would look for the candy where *she* believes it is, not where *they* know it is. But when asked to determine the probability that Sally would look in any one of different locations, even adults think that she is more likely to look in the place *they* know the candy really is, only because they know that. So people have an egocentric tendency in both thinking about other's beliefs and in interpreting what they say; they have experience recovering from that, but they don't always succeed.

The egocentric tendency that we discovered is no small matter. Though adults do better than kids, they still show a surprising disregard for the perspective of the other. Why would adults move the truck when they clearly know that the director could not have known about that particular truck? Whenever adults did this in our experiments, they were unambiguously committing an egocentric error. In fact, the great majority of adult subjects in our experiments (around eighty percent) committed such error at least once during the session (Keysar, Lin, and Barr 2003). And this was not because their private knowledge was more compelling than the knowledge shared with the director. When the hidden truck is smaller than the intended truck, the hidden truck is a better, more compelling referent than the intended one. But this difference was not crucial. Even when the two trucks were of the same size, adults were just as likely to commit the egocentric error (Lin and Keysar 2005). In this case, they tended to ask "which truck," neglecting to use their knowledge that the director could only have meant the one he could see. If asymmetry between the intended and private object cannot explain the egocentric behavior, what can?

3.2. Attention and egocentric understanding

One could explain the egocentric tendency we discovered in at least two ways. First, one's own perspective is dominant and provides a compelling interpretation of what others say. Secondly, the consideration of other's beliefs is *not* automatic. Instead, it is an effortful process; it requires cognitive resources, and is easily disrupted. If this is true, then people's interpretations should depend on the resources available to their working memory. People differ in the capacity of their working memory, and this difference affects performance on a variety of cognitive tasks (Baddley 1986; Just and Carpenter 1992). Typically, performance on tasks that depend on memory capacity deteriorates as working memory capacity decreases. In contrast, automatic processes are unaffected by working memory variations. We compared the performance of people with a high-capacity working memory to those with low capacity in our perspective taking task. Indeed, people with relatively low working memory capacity showed a much stronger egocentric tendency than those with high capacity: They were much more likely to be distracted by the hidden truck (Lin and Keysar 2005).

Variation in capacity determines how much working memory is available to different individuals, but memory resources can also vary as a function of external demands. For instance, a phone conversation while driving could deplete attentional resources, thus leaving the driver less able to respond to unexpected problems (Strayer and Johnson 2001). We manipulated such external "cognitive load" by asking subjects to keep in mind either two (low load) or five (high load) sets of numbers while following instructions. Indeed, with a high external load subjects were much more egocentric than with low external load; they behaved like subjects who have a low working memory capacity. The ability to consider other's beliefs, then, is very vulnerable. It is the first thing that is affected by the lack of mental resources. In contrast, egocentric interpretations are robust and less vulnerable to fluctuations in working memory and resources.

3.3. Attention and non-egocentric understanding: Culturally-induced habits

Our findings that lack of attentional resources makes understanding even more egocentric raises the possibility that focused attention can eliminate the egocentric element from comprehension altogether. We have tried to eliminate it in a variety of ways, by stressing the irrelevance of one's privileged information, by giving feedback over the course of the experiment and so on. While such attempts attenuated the egocentric element, they never eliminated it. We therefore considered the possibility that a much stronger force may be more effective – long-term, ingrained cultural habits.

Cultural psychology documents a systematic difference between individualistic-type cultures and more collectivist-type cultures (e.g., Triandis 1995; Triandis, Bontempo, and Villareal 1988). Individualist cultures, typical of Western countries, tend to engender a more independent self, which is defined in terms of one's wishes, choices and achievements. In contrast, collectivist cultures, typical of East Asian countries tend to engender an interdependent self, which is defined in relation to other relevant individuals (Markus and Kitayama, 1991; Ross, Xun and Wilson 2002; Shweder and Bourne 1984). Members of a collectivist culture, then, have a lot of experience focusing their attention on the other. For instance, Cohen and Gunz (2002) demonstrated that people from an East Asian culture are more likely than Westerners to take an "outside" perspective on themselves, as if seeing themselves from another person's eyes. Such culturally-induced habits, then, could allow listeners to focus attention on the other's perspective, eliminating the egocentric tendency we discovered with our mostly Western subjects.

We tested this idea using the same communication game we described above, but the listeners were students at the University of Chicago who were either native English speakers or native speakers of Mandarin (Wu and Keysar 2007a). They received instructions to "move the block," referring to a mutually visible block. Again, there was another block, which was hidden from the director but clearly visible to the subject. The only difference between the two groups was that the Chinese students received the instructions in Mandarin and the native-English speakers received them in English. The results were stunning. The native-English speakers showed the same egocentric tendency we have seen before: The majority of them were confused at least once during the experiment ("which block?"), and even when they were not explicitly confused they were delayed in finding the intended block. In stark contrast, the Chinese students were almost never confused, and they were not delayed because of the hidden block. They were faster and more effective, as if their attention was so focused on the director that they could "see" the array of objects from her perspective. It seems, then, that members of collectivist cultures focus their attention on the other, allowing them to avoid the egocentric element that members of individualist cultures consistently show when they understand what others say.

3.4. Cooperativeness and assessing mental states

The assumption of cooperativeness in comprehension depends on assessing the mental states of the speaker. But understanding does not seem to be guided by what the speaker knows. Instead, listeners interpret what speakers say from their own perspective. They do consider the mental states of the speaker if they need to correct an error, or when culture provides them with powerful tools to put themselves in the shoes of the speaker.

Perhaps cooperativeness would be more likely to play a role when people converse over time, accumulating shared experiences and establishing common ground (Clark, Schreuder and Buttrick 1983; Clark and Carlson 1981). People tend to converge on similar terminology over time (Krauss and Glucksberg 1977). We may start calling something "the worst bush," and continue to call it that, even when context changes and there is no longer a need to distinguish it from other bushes. When we persist in using the same term, it is as if there is a tacit agreement on the meaning. It seems cooperative because if we change what we call it, it might signal a change in referent (Clark 1987). Brennan and Clark (1996) argued that such cooperativeness is at the heart of people's tendency to use terminology consistently over time. If you call a bush a bush, and then suddenly switch and call it a shrub, people are surprised (Metzing and Brennan 2003). It seems that people establish mutual terminology and expect each other to cooperate and adhere to it.

But listeners' expectations are actually independent of cooperativeness. When people establish with a partner a particular way of calling an object, they expect even a new partner to adhere to that terminology. They know that the new partner is not privy to the tacit agreement they established with someone else to call that thing a bush, but they expect it nonetheless (Barr and Keysar 2002). The expectation to call it a bush, then, could not be based on cooperativeness. The same happens when a partner suddenly switches to "shrub," violating a tacit agreement to call it a bush. Listeners are indeed surprised when that happens, but they are just as surprised if the speaker established the agreement with a different person and then switched to a new term when talking to them, even if the speaker doesn't know that they know about that "agreement" (Shintel and Keysar 2007). Listeners do have expectations that speakers keep using the same term for the same thing, but not because they assume the speakers are cooperative; it is because they assume the speakers are consistent.

People's tendency to converge on the same terminology, then, is not governed by considerations of cooperativeness. People do that regardless of what they believe about the other's knowledge and belief. Most strikingly, people behave the same way even when they can't remember past events at all. Hippocampal amnesiacs who repeatedly converse on a set of objects showed the typical convergence over time on a consistent set of terms, just like non amnesiac controls (Duff et al. 2005). Keeping track of other's beliefs, then, is not necessary in order to explain what looks like a cooperative behavior.

The research I reviewed strongly suggests that people understand language from their own perspective, without much consideration for the mental states of the speaker, except when they need to correct an error or when culture provides help with powerful tools. Such egocentric process could be a systematic cause of misunderstanding and miscommunication— but not necessarily. If speakers assume most of the responsibility for disambiguation, if speakers make sure they tailor what they say to the beliefs, knowledge and expectations of their addressees, then communication will not suffer from the listener's egocentric tendency. Next I will evaluate if speakers attempt to do that.

4.   Speaking egocentrically

It is unrealistic to expect people to speak unambiguously. Sources of ambiguity are so numerous that some ambiguity is virtually guaranteed. But as with any performance, speaking need not be devoid of pitfalls in order to function well. A good enough performance is sufficient (Ferreira, Ferraro, and Bailey 2002). Indeed, speakers have many tools to constrain ambiguity and reduce it to an acceptable level. And they use these tools routinely. For example, "He broke the glass under the table" has at least two syntactic structures. In one case "under the table" is the location of the glass that he broke, and he may have broken it somewhere else. In the other case, "under the table" is where he broke it. To convey only the first meaning, one could explicitly use a relative clause "He broke the glass that is under the table." Tools such as this syntactic one are readily available to speakers. The question is, do they use them to communicate cooperatively?

4.1. Speakers disambiguate their speech for their own benefit

Several studies suggest that though speakers use such tools to disambiguate meaning, they don't do that in the service of cooperation. They do not disambiguate their speech for the benefit of their addressee. Ferrira and Dell (2000) investigated speakers' tendency to disambiguate expressions such as "The woman knew you…" by distinguishing between "The woman knew you when you were a baby" and "The woman knew that you were cute." The only thing that determined their use of the disambiguating cue was its availability in memory. So while speakers were sensitive to how ambiguous what they said sounded to them, they were not sensitive to how ambiguous it was for a particular addressee (See similar findings in Arnold et al. 2004)

Speakers can use different words to communicate more clearly, but they can also say the same thing with a different intonation. Saying "I should apologize" with a stress on "I" means that I should, but with a questioning intonation on the "I" suggests someone else should apologize. How things are said is a powerful tool that affects what meaning is conveyed, but there is little evidence that it is used for the benefit of addressees. For instance, Kraljic and Brennan (2005) showed that while speakers use prosody for disambiguation, they do this whether their addressee needs it or not. They use intonation even when the addressee has sufficient knowledge to understand who should apologize. So speakers disambiguate because it seems better to them, not because they attempt to be cooperative.

Speakers also pronounce words with varying degrees of clarity. When they talk about something for the first time, they pronounce their words more clearly than when they continue to refer to it (Fowler and Housum 1987). This makes sense for communication and is indeed functional for the addressee. When your friend starts gossiping about a new colleague, it is useful that he pronounces her name, Tzimisce, very clearly. When he mentions it again and again, his pronunciation is not as clear any more. Vowels are reduced and he says it faster. This is useful for you, because the first time you hear it is when you need help, when you need it to be very clear. After that, your memory fills in the missing information and you have no difficulty understanding the reduced form. Though this helps the addressee, there is no evidence that speakers do it to be cooperative. They pronounce words clearly initially and less clearly subsequently independently of the needs of their addressee (Bard et al. 2000).

Being informative is a central part of being cooperative. So when my colleague asks me where I live I do not tell him "in Chicago" because this would clearly be under-informative. Indeed, Engelhardt, Bailey and Ferreira (2006) found that speakers avoid being under informative. But they also found that speakers systematically err in the other direction. They tend to be over-informative. This is analogous to answering the question "where do you live" by providing my exact address, when my colleague was just trying to make conversation.

But there are cases when people seem to be perfectly informative. Indeed, when people tell stories they seem to provide information at the "right" level. They are more likely to spell things out precisely when things are not obvious. So when they tell a story about stabbing, they are more likely to mention the instrument when it is an ice pick than when it is a knife. In general, they are more likely to provide information when it is atypical than typical. An ice pick is a relatively rare tool for stabbing, a knife is more common. So it seems that speakers are behaving in line with cooperativeness. They take the knowledge and beliefs of their addressees into account, and use information accordingly. As it turns out, speakers are not really doing this because they are sensitive to the knowledge of their addressees. They are just as likely to provide atypical information when their addressees are uninformed as when their addressees are already informed (Brown and Dell 1987; Dell and Brown 1991). Speakers are less likely to mention typical information not because it is obvious to their addressees, but because it is obvious to them.

4.2. Availability, anchoring, and adjustment when speaking

Availability of information is a powerful determinant of how the mind works (Tversky and Kahnemen 1973). It also seems to play an important role in what information speakers rely on. What determines speakers' behavior is not what they believe to be available to their addressee, but what is available to them. When doctors answer patient's questions they could infer how savvy the patient is about medical issues from the way the patient asks the questions. It makes sense that they would then use technical language if the patient used technical language, but use more everyday language if the question did not include technical terms. This is what Jucks, Bromme and Becker (2005) found. But they also found that the tendency to use technical language was just as high when the patient's question was non-technical, but the medical expert consulted a source that used technical terms. The source made the technical terms available, and so the expert was more likely to use them, even though the patient had no access to that source. Availability of information could make speakers look like they are being cooperative when they are not.

A few studies show that speakers do attempt to take their addressee's mental states into account. When we asked people to identify pictures for addressees, they tended to use shared context more than their own private context. But under pressure to communicate quickly, they were just as likely to rely on private context as on context shared with the addressee (Horton and Keysar 1996). Roßnagel (2000, 2004) found a similar pattern with a different methodology; speakers were less able to tailor their speech to their addressees when they were under cognitive load than when their attentional resources were undisturbed. This suggests that though speakers are fundamentally egocentric when they plan what to say, they monitor and attempt to correct errors to tailor their speech to their addressees. But they anchor on the initial egocentric plan, and when the monitoring process is interrupted, with time pressure or cognitive load, they fall back on purely egocentric speech.

Speakers do not seem to be able to monitor for ambiguity very effectively. A purely linguistic ambiguity is particularly hard to detect. When speakers attempt to identify a picture of a baseball bat for addressees, they often call it a bat, even if this may lead the addressees to select an animal bat. In contrast, it is easier for speakers to avoid referential ambiguity; when two animal bats are present, they often distinguish them by adding an adjective, like "the large bat" (Ferreira, Slevc, and Rogers 2005). Speakers show a similar difficulty with linguistic ambiguity when trying to use intonation to disambiguate syntactically ambiguous sentences. Acoustic analysis shows that though speakers attempt to, they do not include the necessary acoustic cues (Allbritton, McKoon, and Ratcliff 1996).

4.3. Do speakers know when they are unclear? The problem of construal

Speakers' difficulty in disambiguating what they say could lead to misunderstanding, but it doesn't have to. If speakers can gauge their effectiveness, they may be able to anticipate that their addressee would have difficulty understanding them. So speakers need not necessarily be always clear, but the question is, are they calibrated? Can they tell when they conveyed their intention successfully and they when didn't succeed?

We found that speakers are not calibrated. They are systematically biased to think that they are understood when they are not (Keysar and Henly 2002). We asked subjects to say syntactically ambiguous sentences so that another subject will understand them as unambiguous. For instance, they said "Angela killed the man with the gun," trying to convey the idea that Angela used the gun to kill the man, not that he had the gun. Then we asked them which of the two meanings the listener understood, and compared it to the meaning the listener actually understood. Only about 10% were calibrated and a few underestimated. The great majority of speakers tended to overestimate their ability to convey the message. The overestimation was quite dramatic. When speakers thought they were understood, fifty percent of the time they were wrong. One might suspect that such overestimation is exaggerated because of the experimental situation, but it is probably the other way around. In the experiment speakers were provided with both meanings, and actively attempted to disambiguate the sentence. This must have helped them contrast the meaning and exaggerate the one they intended to convey. In a typical conversation speakers do not normally consider alternative meanings to what they say. So in "real life" they may not even realize that there is a need to disambiguate it. This surely would result in an even more dramatic overestimation.

When and why do speakers overestimate their effectiveness? The answer is, under many types of circumstances, and for many reasons. Communication affords a variety of situations that lend themselves to such overestimation. When speakers attempt to use intonation to disambiguate syntactic ambiguity, they use cues. So they would exaggerate the stress on Angela to convey that she was the one who killed him. But they know what they attempt to convey, and they know how they are doing it. This private knowledge makes the stress on Angela sound objectively clear. But it only sounds like that to them, because they already know what they are trying to convey. Such "construal" is fundamental to our interpretive system (Griffin and Ross 1991; Ross 1990) and it introduces a paradox to communication: Because we know what our intention is, our communication seems to convey it uniquely; it seems to have only that meaning. This illusion was demonstrated with non-linguistic communication by having people tap a song so that an audience would be able to identify it. Just like our speakers, tappers greatly overestimated their effectiveness (Newton 1990). Instead of a mental orchestra that accompanies the tapping, our speakers had in mind their intended meaning, which caused them to hear what they said as effective.

This construal problem in communication is very pervasive, making people less calibrated about their effectiveness. For instance, it is easier to communicate on the phone than via email. It is easier to communicate face to face than on the phone. These differences are particularly clear when intonation is important. For instance, people were asked to convey either a sarcastic message or a sincere one, and to estimate which message their addressee understood. Given that a sarcastic tone is much easier to convey in speech, people managed to convey it much more effectively by speaking than via email. However, they thought that they were just as effective in both media (Kruger et al. 2006). Kruger et al. found that people are not sensitive to difficulties that different media introduce and don't appreciate the handicap of lack of intonation in email messages; but even when they can use intonation, they overestimate the effectiveness of those cues (Keysar and Henly 2002). Given that media variations abound and that cues to meaning are of many sorts, speakers have ample opportunity to wrongly conclude that their addressee understood them.

One way that speakers may be cooperative is to actively consider the mental states of their addressees in order to tailor their communication to them. They would evaluate what they say *vis á vis* what they know about what their addressee knows. This might be too daunting a task for the human mind. Instead, speakers may use a rougher heuristic of who knows what. They may not consider if each piece of information is known by the other, but instead keep track of how much information they share with the other. Under some circumstances, this may lead people to miscommunicate more with people who share a lot with them than with people who share little information with them. This is precisely what we found (Wu and Keysar 2007b). The more information people share, the more they tend to confuse their addressee when they communicate over new information. This is particularly pertinent to the possibility of miscommunication because people typically expect the opposite. They expect to communicate better when they share more with others than when they share less.

5.   Conclusion

Listeners rely on their own perspective when they understand language; they do not routinely use knowledge of the speaker's mental states when they understand what the speaker says. They show a fundamental egocentric tendency coupled with an earnest attempt to understand the speaker from his or her own perspective. Assumptions of cooperativeness, then, come into play only as part of a corrective mechanism, if they do at all. Speakers do not seem to be guided by cooperativeness either. They disambiguate what they say, but mainly because it seems ambiguous to them, not because of how ambiguous it is for their addressee.

Egocentric speech and egocentric understanding could introduce a systematic reason for miscommunication. Private knowledge affects processing in two ways. Sometimes it seems to be shared when it is not. With the use of effortful processes one could undo this. The more insidious impact comes from its "construal" effect. Private knowledge can make an ambiguous utterance seem unambiguous by "construing" it. Once it seems unambiguous, it seems objectively unambiguous; it seems independent of the private knowledge that disambiguated it. This is particularly relevant for speakers who are trying to convey an intention, which is always private knowledge, via an utterance, which is always ambiguous. Consequently, speakers have difficulty gauging their ability to convey their message and they systematically overestimate their effectiveness. Therefore, they are less likely to be able to design their utterances for the benefit of their addressee, and less likely to notice when their addressee misunderstands them.

If this is true, then why is communication so successful? Why are people so effective in conveying and understanding intentions? One answer is that successful communication is overdetermined. Even when people are not acting as cooperative agents they may communicate successfully because the context is powerful. The other answer is that we don't know how successful communication really is. It took me two days to figure out why my wife was not talking to me, and it took her two years to agree that one could understand what she said differently from what she meant. Furthermore, much of miscommunication may simply go unnoticed. You may tell a friend you really liked that movie about the journalist from Kazakhstan who is touring the United States, and the friend may think you were being sarcastic. You proceed to talk about other movies without ever knowing that he misunderstood you. By definition, we don't know how often miscommunication goes unnoticed. This cluelessness distorts our performance feedback, making it very difficult to know when we are communicating well and when we are not.

Acknowledgement

The writing of this paper benefited from the support of PHS grant R01 MH49685-06

Note

1. This paper is an updated version of: Keysar, Boaz. 2007. Communication and miscommunication: The role of egocentric processes. *Intercultural Pragmatics* 4 (1): 71-84.

References

Allbritton, David W., Gail McKoon, and Roger Ratcliff
    1996   Reliability of prosodic cues for resolving syntactic ambiguity. *Journal of Experimental Psychology: Learning, Memory, and Cognition* 22: 714-735.
Avis, Jeremy, and Paul L. Harris
    1991   Belief-desire reasoning among Baka children: Evidence for a universal conception of mind. *Child Development* 62: 460–467.
Baddeley, Alan
    1986   *Working Memory*. Oxford: Clarendon Press.
Bard, Ellen G., Anne H. Anderson, Catherine Sotillo, Matthew Aylett, Gwyneth Doherty-Sneddon, and Alison Newlands
    2000   Controlling the intelligibility of referring expressions in dialogue. *Journal of Memory and Language* 42: 1-22.
Birch, Susan, and Paul Bloom
    2007   The curse of knowledge in reasoning about false beliefs. *Psychological Science.*18 (5): 382-386.
Brennan, Susan. E., and Herbert H. Clark
    1996   Conceptual pacts and lexical choice in conversation. *Journal of Experimental Psychology: Learning, Memory, and Cognition* 22: 1482-1493.

Brown, P. Margaret, and Gary S. Dell
  1987   Adapting production to comprehension: The explicit mention of
          instruments. *Cognitive Psychology* 19: 441-472.
Clark, Eve V.
  1987   The principle of contrast: A constraint on language acquisition. In
          Mechanisms of Language Acquisition, Brian MacWhinney (ed.).
          Hillsdale, N. J.: Lawrence Erlbaum Associates.
Clark, Herbert H., and Thomas B. Carlson
  1982   Hearers and speech acts. *Language* 58: 332-373.
Clark, Herbert H., Robert Schreuder, and Samuel Buttrick
  1983   Common ground and the understanding of demonstrative reference.
          Journal of Verbal Learning and Verbal Behavior 22: 245-258.
Cohen, Dov, and Alex Gunz
  2002.   As Seen by the Other…:Perspectives on the Self in the Memories
          and Emotional Perceptions of Easterners and Westerners.
          Psychological Science 13: 55–59.
Dell, Gary S., and P. Margaret Brown
  1991   Mechanisms for listener-adaptation in language production: Limiting
          the role of the "model of the listener." In *Bridges between
          psychology and linguistics*, Donna Jo Napoli and Judy Kegl (eds.),
          105-129. New York: Academic Press.
Duff, Melissa C., Julie Hengst, Daniel Tranel, and Neal J. Cohen
  2005   Development of shared information in communication despite
          hippocampal amnesia. *Nature Neuroscience* 9: 140 – 146.
Engelhardt, Paul E., Karl G.D. Bailey , and Fernanda Ferreira
  2006   Do speakers and listeners observe the maxim of quantity? *Journal of
          Memory and Language* 54: 554-573.
Epley, Nicholas, Boaz Keysar, Leaf VanBoven, and Thomas Gilovich
  2004   Perspective Taking as Egocentric Anchoring and Adjustment.
          Journal of Personality and Social Psychology 87: 327-339.
Ferreira, Fernanda, Vittoria Ferraro, and Karl G.D. Bailey
  2002   Good-enough representations in language comprehension. *Current
          Directions in Psychological Science* 11: 11-15.
Ferreira, Victor S., and Gary S. Dell
  2000   The effect of ambiguity and lexical availability on syntactic and
          lexical production. *Cognitive Psychology* 40: 296-340.
Ferreira, Victor, L. Robert Slevc, and Erin S. Rogers
  2005   How do speakers avoid ambiguous linguistic expressions? *Cognition*
          96: 263-284.
Fodor, Jerry A.
  1985   Precis of the modularity of mind. *The Behavioral and Brain Sciences*
          8: 1-42.

Grice, H. Paul
  1957   Meaning. *Philosophical Review* 66: 377-388.
Griffin, Dale W., and Lee Ross
  1991   Subjective construal, social inference, and human misunderstanding. In *Advances in Experimental Social Psychology* 24, Mark P. Zanna (ed.). New York: Academic Press.
Horton, William S. and Boaz Keysar
  1996   When do speakers take into account common ground? *Cognition* 59: 91-117.
Jucks, Regina, Ranier Bromme, and Bettina-Maria Becker
  In Press   Lexical entrainment – Is Expert's Word Use Adapted to the Addressee? *Discourse Processes*.
Just, Marcel A., and Patricia A. Carpenter
  1992   A capacity theory of comprehension: Individual differences in working memory. *Psychological Review* 99: 122-149.
Keysar, Boaz, and Anne S. Henly
  2002   Speakers' overestimation of their effectiveness. *Psychological Science* 13:207-212.
Keysar, Boaz, Shuhong Lin, and Dale J. Barr
  2003   Limits on theory of mind use in adults. *Cognition* 89: 25-41.
Krauss, Robert M., and Sam Glucksberg. 1977. Social and nonsocial speech. *Scientific American* 236: 100-105.
Kruger, Justin, Nicholas Epley, Jason Parker, and Zhi-Wen Ng
  2006   Egocentrism over email: Can we communicate as well as we think? *Journal of Personality and Social Psychology* 89: 925-936.
Lin, Shuhong, and Boaz Keysar
  2005   The role of attention in perspective taking. *The 45th annual meeting of the Psychonomic Society*: Toronto.
Markus, Hazel, and Shinobu Kitayama
  1991   Culture and the self: Implications for cognition, emotion, and motivation. Psychological Review 98: 224-253.
Metzing, Charles, and Susan E. Brennan
  2003   When conceptual pacts are broken: Partner-specific effects on the comprehension of referring expressions. J*ournal of Memory and Language* 49: 201-213.
Newton, Elizabeth L.
  1990   The rocky road from actions to intentions. Ph.D. diss., Stanford University.
Perner, Joseph
  1991   *Understanding the representational mind.* Cambridge, MA: MIT Press.

Perner, Joseph, Susan Leekam, and Heinz Wimmer
1987   Three-year-olds' difficulty with false belief: The case for a
         conceptual deficit. *British Journal of Developmental Psychology* 5:
         125-137.
Roßnagel, Christian
2000   Cognitive load and perspective-taking: applying the automatic-
         controlled distinction to verbal communication. *European Journal of
         Social Psychology* 30: 429-445.
2004   Lost in thought: Cognitive load and the processing of addressees'
         feedback in verbal   communication. *Experimental Psychology* 51:
         191-200.
Ross, Lee
1990   Recognizing the role of construal processes. In *The legacy of
         Solomon Asch: Essays in cognition and social psychology*, Irvin
         Rock (ed.), Hillsdale, NJ: Erlbaum.
Ross, Michael, W.Q. Elaine Xun, and Anne E. Wilson
2002   Language and the bicultural self. *Personality and Social Psychology
         Bulletin* 28: 1040-1050.
Sabbagh, Mark A., Fen Xu, Stephanie M. Carlson, Louis J. Moses, and Kang Lee
2006   The development of executive functioning and theory of mind.
         Psychological Science 17 (1): 74-81.
Shintel, Hadas, and Boaz Keysar
2007   You Said it Before and You'll Say it Again: Expectations of
         Consistency in Spoken Communication. *Journal of Experimental
         Psychology/ Learning, Memory and Cognition*: 357-369.
Sperber, Dan, and Deirdre Wilson
2002   Pragmatics, modularity and mind-reading. *Mind and Language* 17:
         3-23.
Strayer, David L., and William A. Johnson
2001   Driven to distraction: Dual-task studies of driving and conversing
         on a cellular phone. *Psychological Science* 12: 462-466.
Triandis, Harry C.
1995   *Individualism and collectivism.* Boulder: Westview Press.
Triandis, Harry C., Robert Bontempo, Marcelo J. Villareal, Masaaki Asai, and
         Nydia Lucca
1988   Individualism and collectivism: cross-cultural perspectives on self-
         in-group relationships. *Journal of Personality and Social Psychology*
         54: 323-338.
Tversky, Amos, and Daniel Kahneman
1973   Availability: A heuristic for judging frequency and probability.
         Cognitive Psychology 5: 207-232.

Wellman, Henry M.
   1990   *The child's theory of mind.* Cambridge, MA: MIT Press.
Wellman, Henry M., David Cross, and Julianne Watson
   2001   Meta-analysis of theory-of-mind development: The truth about false
            belief. *Child Development* 72: 655-684.
Wu, Shali and Boaz Keysar
   2007a   Cultural effects on perspective taking. *Psychological Science* 18:
            600-606.
   2007b   The effect of information overlap on communication effectiveness.
            Cognitive Science 31 (1): 169-181.

List of contributors

Stavros Assimakopoulos is a researcher at the University of Edinburgh, where he recently completed his PhD in linguistics. His work revolves around the implications that relevance-theoretic pragmatics carry for the study of the human cognitive capacity for language and his research interests also include cognitive and evolutionary psychology, philosophy of mind, and translation theory.

Herbert L. Colston is Professor and Chair of the Psychology Department at the University of Wisconsin-Parkside. He completed his graduate studies at the University of California, Santa Cruz. He has published numerous books and articles on varieties of topics concerning language and cognition, primarily focusing on comprehension and use of figurative and indirect language.

Nicholas J. Enfield is a staff scientist in the Language and Cognition Group at the Max Planck Institute, Nijmegen. He carries out field-based research on Lao and other languages of mainland Southeast Asia. His theoretical interests include semiotic structure and process, social cognition in human interaction, and micro-macro relations in semiotic systems. Publications include Ethnosyntax (2002, OUP), Linguistic Epidemiology (2003, Routledge), Roots of Human Sociality (2006, Berg, with SC Levinson), Person Reference in Interaction (2007, CUP, with T Stivers), A Grammar of Lao (2007, Mouton) and The Anatomy of Meaning (2008, CUP).

Montserrat González is Assistant Professor of English Linguistics and Discourse Analysis at the Faculty of Translation and Interpretation, Universitat Pompeu Fabra, Barcelona. Her lines of research are in the fields of discourse structure, pragmatics, and text linguistics, with a particular interest in the semantic-pragmatic interface of linguistic devices found in oral discourse. She has carried out contrastive analysis of English and Catalan, focusing on pragmatic markers and epistemic modality features. Her dissertation on pragmatic markers in English and Catalan oral narrative was published by John Benjamins in 2004.

Michael Haugh is a lecturer in the School of Languages and Linguistics at Griffith University, Australia. His main research interests lie in pragmatics and intercultural communication, where he has published work on politeness, face, and implicature in the *Journal of Pragmatics, Intercultural Pragmatics, Multilingua, Journal of Politeness Research,*and *Pragmatics and Discourse*. He has also recently edited a special issue on "Intention in pragmatics" for *Intercultural Pragmatics*, and is co-editing a forthcoming volume *Face, Communication, and Social Interaction* (Equinox, London).

William S. Horton received his Ph.D. in Cognitive Psychology in 1999 from the University of Chicago. He is currently an Assistant Professor in the Department of Psychology at Northwestern University. His research interests include conversational pragmatics, language production, comprehension of non-literal language, and age-related changes in discourse processing.

Kasia M. Jaszczolt is Reader in Linguistics and Philosophy of Language at the University of Cambridge and Fellow of Newnham College, Cambridge. Her books include *Discourse, Beliefs and Intentions* (Elsevier 1999), *Semantics and Pragmatics* (Longman 2002), *Default Semantics* (OUP 2005), and *Representing Time* (OUP, forthcoming in 2008). She is a member of editorial boards of several linguistics journals and book series, including *Journal of Pragmatics* and *Studies in Pragmatics*.

Boaz Keysar is a Professor of Psychology at the University of Chicago and chair of the Cognitive Program. He received his Ph.D. from Princeton University in 1989 and was a visiting scholar at Stanford University before joining the faculty at the University of Chicago in 1991. Professor Keysar's research centers on the psychology of communication, investigating how people communicate and miscommunicate. Professor Keysar's honors and awards include a John Simon Guggenheim Memorial Foundation Fellowship, major Federal research grants, and a Fulbright Scholarship.

Jacob L. Mey is Professor Emeritus of Linguistics at the University of Southern Denmark and Senior Research Associate at the Universidade de Brasília. He has published in the field of pragmatics, intercultural communication, computer-related humanities, and mainstream linguistics. His latest books include the second edition of *Pragmatics: An Introduction* (Blackwell, Oxford 2001) and *As vozes da Sociedade* ('The voices of society'; in Portuguese; Campinas 2004). He is the (co-)founder and Chief Editor of the *Journal of Pragmatics* (since 1977; Elsevier, Oxford).

Montserrat Ribas is Assistant Professor of Catalan Linguistics and Discourse Analysis at the Faculty of Translation and Interpretation, Universitat Pompeu Fabra, Barcelona, Spain. Her primary lines of research are in Critical Discourse Analysis and Cognitive Linguistics. She is particularly concerned with the interaction between discourse, cognition, and social practices. Her latest publications include "Dominant public discourse and social identities" In: *Communicating Ideologies: Language, Discourse and Social* Practice, edited by M. Pütz, J. Neff, and T. Van Dijk. Frankfurt / New York / Paris: Peter Lang, 2004.

Henk Zeevat is a senior lecturer in computational linguistics at the ILLC, University of Amsterdam since 1992, after working on various research projects in Edinburgh and Stuttgart. His main interests are discourse semantics, discourse structure, philosophy of language, natural language processing, and pragmatics. His most recent publications are about presupposition and the application of optimality theory to pragmatics.

Index